Advances in Genome Science

(Volume 4)

Genes in Health and Disease

Editor

Atta-ur-Rahman, *FRS*

Kings College
University of Cambridge
Cambridge
UK

CONTENTS

CHAPTERS

PREFACE

This fourth volume of the Advances in Genome Sciences continues the tradition of the previous volumes in bringing together a range of comprehensive and cutting edge reviews that present the exciting recent advances in this important field. This volume features updates of reviews published earlier in *Current Genomics*. The first 3 chapters cover specific types of human diseases followed by reviews on RNA research, fungal genomics and plant genomics.

The first chapter by Butler on current information about Prader-Willi Syndrome (PWS), a rare trisomy which can lead to childhood obesity. It summarize the clinical features, current understanding of genetic causes and natural history with clinical presentation of individuals suffering from PWS. In the second chapter Franze' *et al* provide an overview of recent developments regarding pathogenesis of syndromic deafness with an emphasis on relevant clinical genotype-phenotype correlations. Effects of some mutations on channels function and the relevant role of the hemichannel are discussed. These genetic defects can result in syndromic skin diseases and deafness. The Italian team led by Antonelli *et al.* discuss the use of tyrosine kinase inhibitors (TKIs) for thyroid cancer therapy. Tyrosine kinase inhibitors (TKIs) inhibit the process of auto-phosphorylation and activation by tyrosine kinases and they are emerging as new therapies against aggressive TC, including differentiated thyroid cancer (DTC), medullary thyroid cancer (MTC) and anaplastic thyroid cancer (ATC).

Chapter 4 by Dostie *et al.* review the known lncRNA transcription regulation activities. They also discuss the potential mechanisms by which ncRNAs might exert three-dimensional transcriptional control and the recent studies that have revealed new facts about their role in shaping the human genome.

The next chapter by Mate and Alcalde comprehensively reviews the progress on genetic engineering of laccase enzymes of *Saccharomyces cerevisiae* and other fungi. Fungal laccases are biocatalysts with a wide range of potential applications that extending from bioremediation to novel green processes. Fuelled by molecular oxygen, they can oxidise many different types of substrates including xenobiotic compounds (pesticides, industrial dyes, PAHs), biopolymers (lignin, starch, cellulose) and other molecules. The authors discuss the efforts to convert fungal laccases into valuable biomolecular platforms on which new functions can be incorporated by directed evolution.

In the next chapter Aceto and Gaudio describe the processes involves in orchid genetics and evolution. This review presents a couple of significant genetic models such as the ABCDE model and MADS-box model which can be used to engineer different varieties of these plants. In chapter 7, Leister *et al.* present a review on the genome of *Arabidopsis thaliana* with particular emphasis on the present status of resources and concepts for systematic studies of gene function in *A. thaliana*. Current perspectives on the utility of loss-of-function and gain-of-function mutants are discussed in light of the genetic and functional redundancy of many *A. thaliana* genes. In the final chapter by Wani *et al.*, the authors review certain advanced applications of plant genetics, *viz.* engineering cold-stress tolerance in plants. Numerous physiological and molecular changes occur during cold acclimatization which indicate that the phenomenon of cold resistance is more complex than perceived and involves a number of pathways. This research area presents an interesting avenue for agriculture in temperate regions where the cold climate limits the production of agricultural products.

I hope that the series of chapters in this volume will stimulate readers to learn more about how the genome of living organisms shapes their existence in our world and how it opens up new possibilities for developing untapped resources for the benefit of mankind.

I would like to express my gratitude to all the authors for their excellent contributions. I would also like to thank the entire team of Bentham Science Publishers, particularly Mr. Omer Shafi (Assistant Manager Publications), Mr. Shehzad Naqvi (Senior Manager Publications) and team leader Mr. Mahmood Alam (Director Publications) for their excellent efforts. We are confident that this volume will receive wide appreciation from students and researchers.

Atta-ur-Rahman, FRS
Honorary Life Fellow
Kings College
University of Cambridge
Cambridge
UK

CONTRIBUTORS

Alessandro Antonelli	Department of Clinical and Experimental Medicine, University of Pisa, Via Savi, 10, 56126, Pisa, Italy
Anja Schneider	Lehrstuhl für Molekularbiologie der Pflanzen (Botanik), Department Biologie I, Ludwig-Maximilians-Universität München, Germany
Annamaria Franzé	CEINGE-Biotecnologie Avanzate, Naples, Italy and Unità di Audiologia, Dipartimento di Neuroscienze, Scienze Riproduttive e Odontostomatologiche, Università di Napoli "Federico II", Naples, Italy
Cordelia Bolle	Lehrstuhl für Molekularbiologie der Pflanzen (Botanik), Department Biologie I, Ludwig-Maximilians-Universität München, Germany
Dario Leister	Lehrstuhl für Molekularbiologie der Pflanzen (Botanik), Department Biologie I, Ludwig-Maximilians-Universität München, Germany
Diana M. Mate	DWI-Leibniz Institute for Interactive Materials, Forckenbeckstraβe 50, 52056 Aachen, Germany
Elio Marciano	Unità di Audiologia, Dipartimento di Neuroscienze, Scienze Riproduttive e Odontostomatologiche, Università di Napoli "Federico II", Naples, Italy
Enke Baldini	Department of Experimental Medicine, "Sapienza" University of Rome, Viale dell'Università 30, 00185 Rome, Italy
Gabriele Materazzi	Department of Surgical, Medical, Molecular Pathology and Critical Area, University of Pisa, Via Savi, 10, 56126, Pisa, Italy
Gulzar Sanghera	RRS, Punjab Agricultural University, Post Box No. 34 Kapurthala, Punjab, 144601, India
Jennifer Lynn Crutchley	Department of Biochemistry, and Goodman Cancer Research Center, McGill University, 3655 Promenade Sir-William-Osler, Room 815A, Montréal, Québec, H3G1Y6, Canada
Josée Dostie	Department of Biochemistry, and Goodman Cancer Research Center, McGill University, 3655 Promenade Sir-William-Osler, Room 815A, Montréal, Québec, H3G1Y6, Canada
Luciano Gaudio	Department of Biology, University of Naples Federico II, Napoli, Italy
Merlin G. Butler	Departments of Psychiatry & Behavioral Sciences and Pediatrics, University of Kansas Medical Center, Kansas City, Kansas, USA

Miguel Alcalde	Department of Biocatalysis, Institute of Catalysis, CSIC, Cantoblanco, 28049 Madrid, Spain
N.B. Singh	Department of, Plant Breeding and Genetics, COA, CAU, Imphal, 795001, Manipur India
Paolo Miccoli	Department of Surgical, Medical, Molecular Pathology and Critical Area, University of Pisa, Via Savi, 10, 56126, Pisa, Italy
Poupak Fallahi	Department of Clinical and Experimental Medicine, University of Pisa, Via Savi, 10, 56126, Pisa, Italy
Salvatore Ulisse	Department of Experimental Medicine, "Sapienza" University of Rome, Viale dell'Università 30, 00185 Rome, Italy
Sandra Iossa	CEINGE-Biotecnologie Avanzate, Naples, Italy and Istituto di Genetica e Biofisica "A. Buzzati Traverso", C.N.R., Naples, Italy
Saroj Kumar Sah	Department of Biochemistry, Molecular Biology, Entomology and Plant Pathology, Mississippi State University, MS-39762, USA
Serena Aceto	Department of Biology, University of Naples Federico II, Napoli, Italy
Shabir H. Wani	Division of Genetics and Plant Breeding, SKUAST-K, Srinagar-190025, Jammu and Kashmir, India
Silvia Martina Ferraril	Department of Clinical and Experimental Medicine, University of Pisa, Via Savi, 10, 56126, Pisa, Italy
Waseem Hussain	Department of Agronomy & Horticulture. University of Nebraska, Lincoln, 68583-0915, USA
Xue Qing David Wang	Department of Biochemistry, and Goodman Cancer Research Center, McGill University, 3655 Promenade Sir-William-Osler, Room 815A, Montréal, Québec, H3G1Y6, Canada

Advances in Genome Science

(Volume 4)

Genes in Health and Disease

CHAPTER 1

Prader-Willi Syndrome: A Rare Obesity-related Genomic Imprinting Disorder

Merlin G. Butler*

Departments of Psychiatry & Behavioral Sciences and Pediatrics, University of Kansas Medical Center, Kansas City, Kansas, USA

Abstract: Prader-Willi syndrome (PWS) is a complex genetic disorder that affects multiple systems due to genomic imprinting errors in which loss of paternally expressed genes from the chromosome 15q11-q13 region causes this rare obesity related disorder. About 70% of individuals with PWS will have a paternal *de novo* deletion of the 15q11-q13 region consisting of two subtypes (*i.e.*, larger Type I or smaller Type II). A second genetic cause is maternal disomy 15 in which both 15s are from the mother and seen in about 25% of cases. The remaining subjects have either defects in the imprinting center that controls the activity of genes under it's control. This syndrome is characterized by a typical facial appearance, hypotonia with a poor suck and feeding difficulties during infancy, hypogonadism and hypogenitalism, short stature and growth and other hormone deficiencies with short stature and small hands and feet. Learning and behavioral problems (*e.g.*, skin picking, temper tantrums) are present in the majority of subjects. Food seeking and hyperphagia leads to obesity in early childhood. Obesity is a significant health problem and PWS is considered the most common known genetic cause of life-threatening obesity in children. Growth hormone therapy is often prescribed to improve stature, body composition (increased muscle mass and strength with lowered fat quanity). Syndrome specific standardized growth charts are now available to assist in monitoring growth patterns during growth hormone treatment from infancy to adulthood. The chromosome 15q11-q13 region contains multiple genes and transcripts in which a dozen are imprinted and paternally expressed and when disturbed leads to PWS. Angelman syndrome, an entirely different disorder is due to loss of a maternally expressed gene (*i.e.*, UBE3A) located in the same chromosome region. Other genes in the area are either bialletically (normally) expressed or show paternal bias of expression. This review will summarize the clinical features, current understanding of genetic causes and natural history with clinical presentation of individuals with PWS.

Keywords: Angelman syndrome, clinical presentation and differences, deletion, genomic imprinting, genotype/phenotype, growth hormone deficiency, maternal disomy 15, obesity, paternal expression, Prader-Willi syndrome, PWS genetic subtypes.

***Corresponding author Merlin G. Butler:** Departments of Psychiatry & Behavioral Sciences and Pediatrics, University of Kansas Medical Center, Kansas City, Kansas, USA; Tel: (913) 588 1800; Fax: (913) 588 1305; E-mail: mbutler4@kumc.edu

Atta-ur-Rahman (Ed.)

INTRODUCTION AND HISTORICAL REVIEW

Prader-Willi syndrome (PWS) is a neurodevelopmental disorder due to loss of imprinted genes that are paternally expressed on chromosome 15q11-q13 region and presenting with a range of behavioral and physical findings. Obesity is the most significant health concern and can be life-threatening, if not controlled [1-3]. About 350,000–400,000 people are affected worldwide with PWS and 17,000–22,000 individuals live in the United States [4]. The incidence is estimated at 1 in 10,000 to 30,000 individuals [1-5]. PWS presents in all races and ethnic groups [2, 3, 6, 7] and most cases are sporadic. In some families, a defective imprinting center is noted which controls differentially expressed genes from the chromosome 15q11-q13 region and causes PWS by errors in processing of genomic imprints in the PWS child carried by the father and inherited from his mother. This error can lead to recurrence of another child with PWS in the same family with a risk as high as 50% [8, 9]. Hence, PWS arises from lack of expression of paternally inherited genes known to be imprinted and located in the 15q11-q13 region.

Approximately 1% of mammalian genes are imprinted and thought to play a role in growth and development. Genomic imprinting is an epigenetic phenomenon whereby the phenotype is modified depending on the sex of the parent contributing the gene allele [10] and arises from epigenetic changes where expression of genes is controlled without changing the DNA code or sequence. This phenomenon is reversible in gametogenesis. Imprinted gene expression is usually regulated through DNA methylation dependent on the parent of origin with mono-allelic gene expression of either the maternal or paternal allele for a particular imprinted locus or gene.

Approximately 70% of individuals with PWS are found to have a non-inherited deletion in the paternally derived chromosome 15q11-q13 region while 25% of individuals result from maternal disomy 15 or both chromosome 15s inherited from the mother with no paternal chromosome 15. A small percent of PWS cases will have genomic imprinting defects due to microdeletions or epimutations of the imprinting center located in the 15q11-q13 region or other cytogenetic rearrangements [1-3, 7, 8, 11]. PWS and Angelman syndrome, an entirely different clinical disorder, are both involved with the same chromosome 15q11-q13 region. They are recognized as the first examples in humans of errors in genomic imprinting [12, 13]. Errors in genomic imprinting are now known to cause other genetic disorders such as Beckwith-Weidemann syndrome [14].

The primary features of PWS include hypotonia and feeding difficulties in infancy when characterized by a poor suck, hypogonadism and hypogenitalism in both males and females. Later food seeking and hyperphagia with onset of obesity in early childhood are noted and short stature with small hands and feet due to growth hormone deficiency. Mild learning (average IQ of 65) and behavioral (skin picking, temper tantrums, stubbornness) problems occur. A particular facial appearance with a small upturned nose, narrow bifrontal diameter and almond-shaped eyes, down-turned corners of the mouth, sticky salivary secretions and lighter skin, hair and eye color are often present [1-3, 7, 15, 16]. The features of PWS were first recognized in 1956 by Prader, Labhart, and Willi [17].

In 1981, Ledbetter *et al.* [18] used high resolution chromosome analysis and showed that more than half of the individuals with PWS had an interstitial deletion of the proximal long arm of chromosome 15 at the q11– q13 region. Butler and Palmer [19] then studied specific staining polymorphisms on chromosome 15 and reported that the chromosome 15 deletion was not inherited but was donated by the father. Those individuals studied were later confirmed using molecular genetic techniques and chromosome 15 DNA markers. Subsequently, a maternal deletion of the 15q11-q13 region was reported in a separate clinical condition, Angelman syndrome. Butler *et al.* in 1986 [16] then reported clinical differences in PWS individuals with and without the chromosome 15q11-q13 deletion, particularly hypopigmentation and greater homogeneity in clinical presentation in those with the deletion.

Nicholls, Butler, *et al.* reported in 1989 [12] that PWS individuals with normal-appearing chromosomes inherited both chromosome 15s from the mother and none from the father by using Southern hybridization of polymorphic DNA markers isolated from the 15q11–q13 region, This unexpected observation was referred to as maternal uniparental disomy 15 or maternal UPD 15. Butler [15] subsequently characterized the hypopigmentation status seen in the majority of individuals with PWS and the 15q deletion and due to loss of the P (pigment) gene located in the 15q11-q13 region. In 2004, Butler *et al.* [20] reported clinical differences between individuals with PWS grouped by two types of the 15q11-q13 deletions; those with the larger typical 15q11-q13 deletion, referred to as Type I involving chromosome 15q breakpoints BP1 and BP3 and those with the smaller Type II deletion involving breakpoints BP2 and BP3. Individuals with the larger 15q11-q13 Type I deletion and PWS were reported with more obsessive compulsions, self-injury (*i.e.*, skin-picking), visual processing deficits and lower measures of academic performance than those PWS individuals with the smaller

15q11-q13 Type II deletion having the four genes intact between BP1 and BP2 that vary in size involving 15q11-q13 region [20, 21]. About 5% of PWS individuals have other atypical 15q11-q13 deletions [2, 3, 11, 22,] and often present with more unusual clinical presentation [23, 24].

Four genes located in the genomic area between proximal 15q11-q13 breakpoints BP1 and BP2 include GCP5, CYFIP1, NIPA1 and NIPA2 and are overly expressed in the brain. When these genes are disturbed or mutated such as the NIPA1 gene then spastic paraplegia or other neurological disturbances can result [20, 25]. Those individuals with PWS having the smaller 15q11-q13 Type II deletion and these four genes intact are reported with fewer behavioral problems. Individuals with only a deletion involving the region between breakpoints BP1 and BP2, the chromosome 15q11.2 BP1-BP2 microdeletion or Burnside-Butler syndrome do not have PWS but have neurological dysfunction including developmental and speech delay or autism [26-28].

CLINICAL STAGES, NATURAL HISTORY AND NUTRITIONAL PHASES

The clinical course of PWS has been described or catorgorized into two distinct stages with early failure-to-thrive in the first stage during infancy and in the second stage in early childhood with hyperphagia and subsequent obesity. With earlier diagnosis and growth hormone treatment in infancy, the clinical course and development of this syndrome has changed. The initial phase of clinical course development is unchanged as it starts in pregnancy with decreased fetal activity followed by central hypotonia, a weak cry and poor suck, a narrow forehead, developmental delay, temperature instability, sticky salivary secretions, hypogonadism and hypogenitalism in infancy followed by feeding problems which may require naso-gastric or gastrostomy tube placement during the first month of life [1-3, 7, 16]. Classically, two clinical phases have been described over time in PWS and now influenced by growth hormone treatment which is common in PWS including in early infancy. Recently, a total of seven different clinical or nutritional phases have been described with five main phases and sub-phases in phases 1 and 2 [29]. Increase in appetite is now described as phase 2b at age 4.5 - 8 years with classical hyperphagia seen in PWS becoming evident during phase 3.

For the non-growth hormone treated PWS child, historically the second stage of clinical course development began around 2 years of age [16] and characterized by developmental delay and onset of hyperphagia leading to obesity if not controlled. Food foraging or seeking, unmotivated sleepiness and physical inactivity are common along with speech articulation problems during this time. Decreased pain sensitivity, skin picking, hypothermia, strabismus, hypopigmentation, scoliosis, sleep apnea, sticky saliva and dental anomalies are common. Often PWS children are easy-going and affectionate, but personality or behavioral problems may develop between 3 and 5 years of age. These include temper tantrums, depression, stubbornness and obsessive compulsions. Behavioral changes are often initiated by withholding of food. Poor peer interactions, inappropriate social behavior and immaturity may also occur at this time into adolescence and adulthood [1, 7] (see Fig. **1**).

Infancy

Infants with PWS can sit independently by 1 year of age with crawling at 16 months, walking at 2 years and talking (10 words) at 39 months [7, 16]. Neonates with PWS are profoundly floppy or hypotonic although no brain anomalies are generally present. Severe hypotonia and decreased muscle mass can lead to respiratory distress and possible asphyxia during illnesses and should be monitored closely. Central adrenal insufficiency has also been reported in PWS and should be ruled out by endrocrine testing [30]. Central apneas are more common in infants studied with PWS [2] but later symptomatic narcolepsy, with or without cataplexy has also been reported in up to 36% of children with PWS [31]. Sleep disruption and sleep disordered breathing have been linked to significant deficits in neurocognitive function, including poor focus, excessive daytime sleepiness and irritability in PWS and in the general population [32, 33].

PWS infants have a weak or absent cry with little spontaneous movement. Hyporeflexia, excessive sleepiness, and poor feeding due to diminished swallowing and sucking reflexes often necessitate gavage feedings and use of special nipples which last for several months. Feeding problems and supplying adequate nutrition for growth and development needs to be addressed. Growth parameters should be regularly assessed (*e.g*., weekly) using recently published growth charts for non- growth hormone treated infants with PWS [34] during the first 6 months and monthly until 2 years of age. Calories are adjusted accordingly, but fats should not be restricted even though the non-growth hormone treated infant with PWS requires less than the recommended allowance (often 60 to 80% normal) to avoid rapid weight gain once the failure-to-thrive stage has passed. A diet consisting of about 30% protein, about 30% fat and about 40% carbohydrates should be in place [29, 35] including vitamin and mineral intake (*e.g*., calcium) monitored by a dietitian during infancy. Early stimulation programs are also recommended, including occupational and physical therapies.

The use of growth hormone (GH) will alter the clinical presentation [7, 36]; thus, syndrome specific growth charts which are now available for PWS infants should be used to monitor growth. Recently, growth hormone receptor gene (GHR) studies in PWS infants and children treated with GH found a higher growth rate (1.7 times) when the GHR gene exon-3 deletion or d3 allele polymorphism is present and seen in 50% of Caucasian controls [37]. These data suggest that those with PWS and having the polymorphic d3 allele may be more sensitive to GH therapy impacting their growth response to GH but more testing in PWS is needed before application with medical management and care.

Childhood

The feeding pattern in PWS may change radically by 18 months of age. An insatiable appetite may present leading to onset of early childhood obesity which needs to be closely monitored. Hyperphagia is a major finding in PWS and typically lasts over the lifetime of an individual with PWS. No known pharmacological agent has been effective in treating hyperphagia once it develops but several clinical trials are underway to investigate treatments for hyperphagia and subsequent obesity.

Global developmental delays and behavior patterns such as temper tantrums, manipulative behavior, difficulty in changing routines, stubbornness and obsessive- compulsions may become more apparent during childhood. Stealing, lying and aggressive behavior are common during this time. PWS children may be affectionate, but are less agreeable and less open to the introduction of new ideas on changes. Without growth hormone treatment, they are generally less physically active than other children [7].

Endocrine abnormalities such as hypothyroidism is seen in about 10% of PWS children while growth and sex hormone deficiency are present in most children with PWS. Adrenal problems are seen in less than 10% of PWS children and usually recognized early [7]. Enamel hypoplasia and dental caries are frequently seen as well as strabismus and myopia with impaired stereoscopic vision; the latter is more common in those with maternal disomy 15 [38]. Hypopigmentation along with small hands and feet and almond-shaped eyes in both sexes becomes more pronounced particularly during mid-childhood. A characteristic body habitus or posture is more noticeable during early childhood with sloping shoulders, central obesity, straight lower leg borders and straight ulnar borders to the hands. Many of these features improve with growth hormone treatment.

Type 2 diabetes mellitus (T2 DM) may be present in PWS and generally results from insulin resistance secondary to the level of obesity and lowered once weight is reduced. However, the study of 74 children with PWS at a median age of 10.2 years showed that none had T2 DM and only 4% had impaired oral glucose tolerance [39]. T2 DM should be managed in PWS with special attention to those children on GH treatment with a higher risk for insulin resistance. Periodic fasting serum glucose and insulin levels are recommended before and after initiation of GH treatment in PWS. Metabolic profiling in PWS was recently reviewed by Irizarry *et al.* [40] and compared with nonsyndromic obese children with analysis of sex differences and role of growth hormone.

Academic achievement is usually impaired during the first 6 years of life although about one-third of children function in the low-normal intellectual range (70–100 IQ). The remaining function in the mild-to-moderate range (50–70 IQ) with an average IQ of 65. There are reported differences in behavior, academic achievement, and cognition between those with the chromosome 15q deletion *versus* those with maternal disomy [41] which will be discussed later.

Many children begin school in mainstream settings, but the intellectual impairment and potential behavioral problems will present difficulties in progressing through regular classroom settings. Special education and support services are often required. Children with PWS have relatively strong reading, visual, spatial, and long-term memory skills, but have weaker math, sequential processing, and short-term memory skills. Verbal skills may be relative strengths, particularly in those children with maternal disomy, but speech articulation can be a problem requiring speech therapy. A common occurrence in those with the typical deletion is an unusual skill of working with jigsaw puzzles [42].

Adolescence and Adulthood

Puberty is absent, delayed or incomplete in both males and females with PWS and infertility is present in the vast majority. Menarche may occur but may not be present until 30 years of age. Amenorrhea or oligomenorrhea is usually present and gonadotropin hormone production is low. Adolescents and young adults with PWS usually look younger than their chronological age [43].

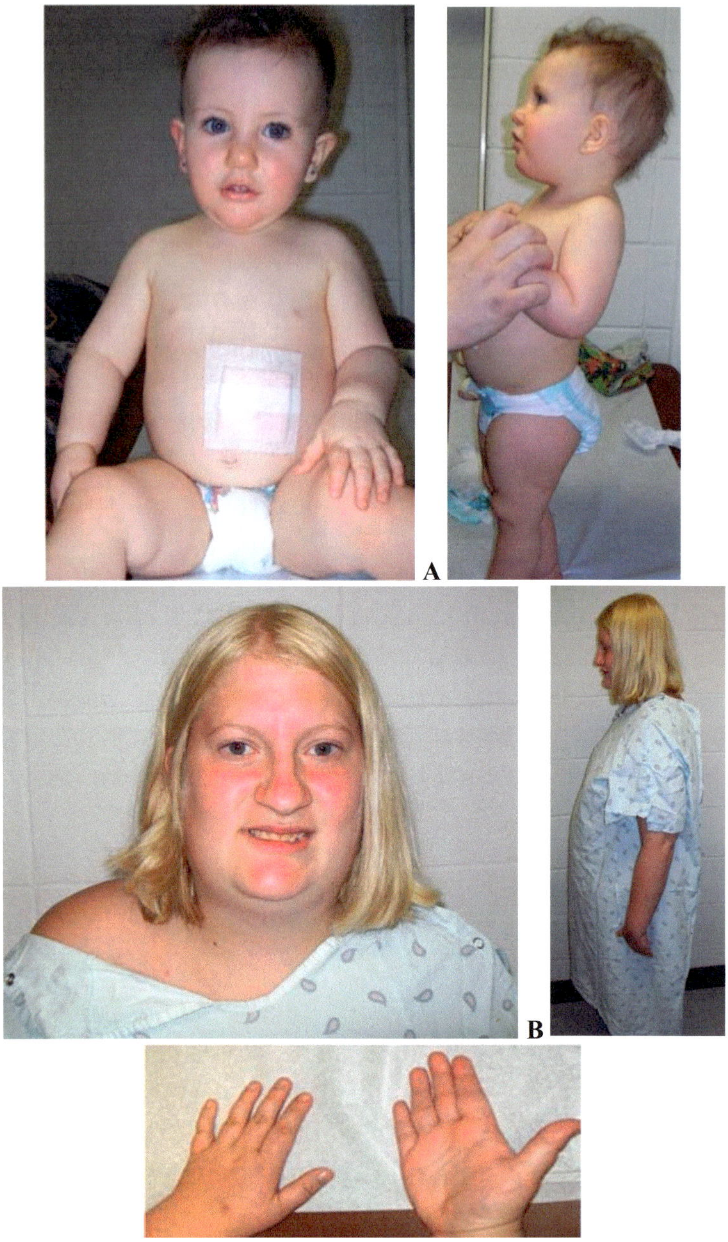

Fig. (1). (**A**). Frontal and profile views of a 16 month old female with Prader-Willi syndrome due to maternal disomy 15. Note the typical facial features of a narrow forehead, short-upturned nose and downturned corners of the mouth. The gastrostomy site is noted along with central obesity. (**B**). Facial, profile and hand views of an 18 year old female with Prader-Willi syndrome due to the typical 15q11-q13 deletion. Note the almond-shaped eyes, a narrow forehead, thin upper lip, hypopigmentation, central obesity and small hands.

Hypogonadism and hypogenitalism occur in the vast majority of individuals with PWS during adolescence. Hypothalamic hypogonadism leads to the abnormal gonadal development and function but recent evidence supports a primary defect in gonadal function, as well. Testosterone and estrogen levels are generally low. Cryptorchidism is present in 80-90% of males with a small penis and hypoplastic scrotum. A hypoplastic labia majora and minora and a small clitoris are seen in most females. Penile size increases modestly in many males during the third or fourth decade of life, but testicular size remains small. Treatment of the small penis with topical or parenteral testosterone has been effective, but normalization of genital development is rare. Testes may descend spontaneously in males during childhood and puberty, but often surgical intervention is required. Precocious development of pubic and axillary hair occurs in about 20% of individuals in both sexes as a consequence of premature adrenarche. Beard and body hair are variable, occurring later than normal, but generally sparse. Inadequate voice change occurs in most males. Little information is known about sexual activity in PWS, although individuals with PWS show an interest in affection and establishing relationships. In PWS females, breast development is normal in about one-half of cases with onset between 9 and 13 years of age. Pubic hair is normal in about 40% of females. Rarely, pregnancies in females with PWS have occurred, but no reported pregnancies produced from PWS males [7, 44].

Approximately 90% of individuals with PWS without growth hormone treatment will have short stature by adulthood. The average untreated adult male is 155 cm (61 inches) tall and the adult female averages 147 cm (58 inches) [7]. Growth standards for untreated males and females with PWS have been reported between 2 to 22 years of age [45]. Growth pattern analysis shows that the 50th centile for height in PWS individuals generally falls below the 5th centile for normal individuals by age 12 to 14 years. Height velocity often declines relative to normal due to the lack of a pubertal growth spurt and growth and sex hormone deficiencies. Inverse correlations between Z-scores for linear measurements (*e.g.*, height, hand and foot lengths) and age indicate a deceleration of linear growth with increasing age relative to normal individuals. Short stature is almost always present by the second decade of life without growth hormone replacement [7, 46-49]. More recent syndrome specific growth charts for non-growth treated PWS males and females between 3 and 18 years of age have been reported [50].

The beneficial effect of GH therapy in childhood into adulthood after GH is discontinued is unclear. A report of GH treatment discontinuation after 12 and 24 months in 14 individuals with PWS revealed an increase in BMI-SDS ($P = 0.008$

and P = 0.003) and visceral fat (P = 0.062 and P = 0.125), respectively [51]. In one study however, improved mean BMI (32.4 vs. 41.2), body composition, lower mean hemoglobin A1c, lower mean insulin resistance, and less hypertension were reported in 20 adults [mean age 25.4 years] with PWS at 7.0 ± 4.4 years after discontinuing treatment initiated at age 11.8 ± 2.7 years when compared with 40 untreated PWS adult group [52]. Butler *et al.* [53] report on a cohort of PWS adults (average age = 32 years) with GH treatment over a two-year period during the first year only. They found that total lean muscle mass and moderate-vigorous physical activity and plasma IGF-1 and high density lipid (HDL) levels were significantly increased while on GH while percent body fat decreased during the 12 months of treatment. IGF-1 and HDL levels returned to near baseline and body fat increased after GH treatment during the second year. These data indicates that beneficial effects of GH are still present even after the epiphyses are fused and long-term GH treatment, in addition to strict diet and exercise program may be necessary to maintain good body composition.

Benefits have been reported with increased lean body mass and decreased body fat mass after 6 to 12 months GH treatment in adults [53-57]; however, GH is not FDA approved for adult PWS individuals unless confirmed by standard adult GH stimulation testing. The prevalence of GH deficiency in adults with PWS ranges from 15% to 95%, depending on the agents used for stimulation testing and the threshold GH level used to define deficiency [54, 55]. A normal IGF-1 level does not exclude the diagnosis of GHD and provocative testing is mandatory to make the diagnosis of adult GH deficiency (AGHD). The Endocrine Society recommends that the insulin tolerance test and the GHRH- arginine test are sufficiently sensitive and specific to establish the diagnosis of AGHD. Glucagon stimulation test can also be used when GHRH is not available and performance of an ITT is either contraindicated or not practical in a given patient [58]. It is recommended that GH dosing regimens be individualized rather than weight-based and to start with low doses (0.1 to 0.2 mg) then titrated according to clinical response, side effects and IGF-1 levels.

Acromicria (small hands and feet) seen during infancy and childhood is more pronounced during adolescence and adulthood. Foot length tends to be more affected than hand length. Scoliosis may also become more pronounced during adolescence and kyphosis may be present by early adulthood. Genu valgus positioning at the knees may also occur [1]. Without intervention, adolescents with PWS may weigh 250 to 300 pounds by their late teens. Beside obesity related complications impacting morbidity and mortality, eating related fatalities do occur, including choking on gorged food and gastric necrosis and rupture [59]. By late adolescence, food stealing and hoarding can be problematic as well as consuming discarded or inedible food items (*e.g.*, frozen food). This often necessitates locking the refrigerator and food cabinets to prevent excessive eating. The eating behavior and complications of obesity can reduce the life expectancy in PWS. Type 2 diabetes mellitus is seen in about 25% of obese adults with PWS. Cognitive impairments preclude adult independent living arrangements. Behavioral and psychiatric problems often require medical treatment and behavioral management. Psychosis is evident by young adulthood in about 10% of individuals with PWS and those with maternal disomy 15 are at a greater risk for psychosis and autism [60, 61]. Caloric diet restriction is lifelong and important to control obesity. Based on population studies, the death rate in PWS is estimated at 3% per year [62].

The decreased muscle tone and mass in PWS contributes to a lower metabolic rate and physical inactivity with subsequent obesity in adolescents and adults. Improvement is seen with growth hormone therapy. Sleep disorders and respiratory dysfunction in PWS such as hypoventilation and oxygen desaturation are also common from childhood to adulthood and need to be closely monitored before growth hormone treatment. Adolescents have a tendency to fall asleep during the day, particularly when they are inactive, and they do not sleep soundly at night. Excessive daytime sleepiness is reported in 70-100% of PWS adults. Hypocretin containing neurons, located in the hypothalamus is thought to play an important role in maintaining wakefulness as well as eating behavior. Congenital dysfunction/developmental failure of the hypocretin system has also been proposed as a potential etiology in this population [63-65]. In addition, the hypocretin (HCRT) gene showed elevated expression patterns in PWS males in genome wide expression studies reported by Bittel *et al.* [66]. Excessive daytime sleepiness (EDS) is a very common feature in PWS, occurring in 70-100% of adults with PWS [7].

Along with daytime hypersomnolence, polysomnographic features of sleep disordered breathing with obstructive, central and mixed sleep apneas are commonly reported in PWS with potential contributing factors of related brain abnormalities, craniofacial problems, hypotonia, obesity and chest body wall deformities. Hypocretin containing neurons, located in the hypothalamus is thought to play an important role in maintaining wakefulness as well as eating behavior. Congenital dysfunction/developmental failure of the hypocretin system has also been proposed as a potential etiology in this population [63-65]. In addition, the hypocretin (HCRT) gene showed elevated expression patterns in PWS males in genome wide expression studies reported by Bittel *et al.* [66].

Weight and behavior control problems are characteristic of PWS and require involvement of the patient, family members and care providers to address the issues. Interventions to control food consumption include locking refrigerators, limiting the amount of food at home, continual supervision during times of food-related events and providing non-food-related rewards. Strict mealtime regimes are important. Individually tailored exercise programs are also encouraged and should be monitored to reduce weight. Adequate protein during times of caloric restriction should be encouraged to help conserve lean body mass and is particularly needed when growth hormone is prescribed to increase stature and muscle mass. Restricted caloric intake with vitamin and calcium supplementation is generally required by age 2 years and beyond to minimize excessive weight gain and prevent osteoporosis [35, 67].

Historically, weight maintenance in non-growth hormone treated children with PWS requires caloric intake of 8-11 kcal per cm of height per day (by contrast, children without PWS require 11-14 kcal per cm per day for adequate growth) [7, 35]. Weight loss has been documented in children by restricting their caloric intake to 7 kcal per cm per day. Similar guidelines have not yet been developed specifically for adolescents and adults with PWS. Hence, a lifetime commitment and close observation are required by all involved in the care of individuals with PWS, regardless of their age or living arrangement. The obesity associated with PWS results from a chronic imbalance between energy intake and expenditure due to hyperphagia, decreased physical activity, reduced metabolic rate and an inability to vomit. Individuals with PWS have a lower lean body mass compared with controls contributing to reduced energy expenditure. Studies have been conducted to determine the relationship among body composition, activity levels and metabolic rates in PWS. Butler and colleagues [68] demonstrated that individuals with PWS had reduced total energy expenditure and a lower lean body mass compared with obese subjects. In addition, resting energy expenditure was significantly reduced by 16% in the individuals with PWS compared with control subjects.

Theodoro and colleagues [69] studied body composition in PWS individuals between 10 to 49 years. Regional fat and lean mass patterns were characterized and significant differences were found between PWS and obese subjects for lean measures of the arms, legs, and trunk. Total lean mass was significantly lower in PWS than in obese subjects for arms, trunk, and legs, demonstrating an unusual body composition and fatness patterns in PWS characterized by reduced lean tissue and increased adiposity.

Because hyperphagia is the most striking behavior in PWS, Holsen and colleagues [70] used functional magnetic resonance imaging (fMRI) to study the brain's response to food stimulation images before and after eating in individuals with and without PWS. When viewing these images before and after eating, the fMRI scans showed that PWS subjects exhibited greater brain activation to food pictures in the post-meal state compared to controls. The activated brain areas included the orbitofrontal cortex, medial prefrontal cortex, insula, hippocampus, and parahippocampal gyrus; several of these regions function to drive eating behavior. The last two decades have seen significant increases in the understanding of mechanisms controlling appetitive behavior, body composition and energy expenditure. Many regions throughout the central nervous system, such as the hypothalamus, play critical roles in these processes by receiving and orchestrating a variety of signals to bring about coordinated changes in energy balance. Ghrelin, a 28 amino acid peptide produced in the stomach, is the only peripheral hormone to transmit a satiety signal. Plasma ghrelin level is higher in obese PWS individuals than in other forms of obesity. It is considered one of the contributing factors in obesity [71, 72]. Circulating ghrelin levels are elevated in young children with PWS before the onset of hyperphagia [73, 74]. It seems unlikely that high ghrelin levels alone are directly responsible for the switch to the hyperphagic nutritional phases in PWS but more studies are needed.

Behavioral and learning problems become more prominent during the teenage years, particularly temper tantrums, self-injury and obsessions. Typical adolescent rebelliousness is often exaggerated, particularly over access to food. Self-injurious behavior is reported in 60% of PWS adolescents and 81% of adults, with skin picking being the most common [75, 76]. Compulsive symptoms have been reported in about 60% of individuals with PWS [77] and occasionally these behaviors worsen during adulthood. In addition, acute psychosis can be seen in about 10% of young adults with PWS, particularly in those with maternal disomy.

Psychotropic agents can be helpful in controlling abnormal behavior, but no specific medication has been universally effective in controlling abnormal behavior or food-seeking behavior. Serotonin agonists have been used to reduce behavioral problems, including temper tantrums and compulsions. However, children and adults with PWS are typically affectionate and outgoing and show a willingness to please others and seek positive attention. When an adult with PWS is surrounded by trained and experienced caregivers, behaviors can be managed.

Adults with PWS have life-oriented goals similar to others entering adulthood, including vocation, independent living and long term relationships. For most persons with PWS, formal education ends between ages 18 and 21. However, a gap often exists between completion of school and entrance into a job-training program. Challenges may be numerous for the adult with PWS and the family members or caregivers. In view of potential health problems such as behavior and hyperphagia, an appropriate environment requires supervised living arrangements with food restrictions. Professional sources knowledgeable about PWS are required for development of training and sheltered employment and housing. Decisions regarding living arrangements and availability of food are usually made unwisely by persons with PWS; therefore it is optimal to have legal guardianship assigned to a parent or other adult.

PRADER-WILLI SYNDROME: GENETICS

Prader-Willi syndrome is a genomic imprinting disorder due to an epigenetic phenomenon that evolved about 150 million years ago and involves modification of the phenotype of an individual depending on the parent of origin. About 1% of mammalian genes are imprinted and frequently affect growth, development and viability. PWS and other obesity-related and overgrowth disorders such as Albright hereditary osteodystrophy and Beckwith-Weidemann syndrome are examples of errors in genomic imprinting [14].

Epigenetics refers to heritable but reversible regulation of various genetic functions, including gene expression, which is influenced by environmental factors. Nearly all imprinted genes have a CpG-rich differentially methylated region which usually relates to allele repression. Methylation of cytokine bases occurring in the CpG dinucleotides involve key regulatory elements of the genes. Defects in the imprinting center controlling the activity of imprinted genes originate from either parent and can lead to different clinical syndromes. For example, the loss of paternally expressed genes from the 15q11-q13 region leads to PWS with the maternal contribution of this chromosome region normally silenced by epigenetic factors (*e.g.*, DNA methylation). The sister genomic imprinting syndrome, Angelman syndrome, is due to loss of maternal expression of genes in the same region that are normally silenced on the paternal chromosome 15.

The chromosome 15q11-q13 region contains about 8 million DNA base pairs including a large cluster of imprinted genes and also a non-imprinted domain in which genes are expressed equally on either the maternal or paternal chromosome

15, but with a few genes showing a paternal expression bias [78, 79]. Novel low copy repeat DNA sequences are clustered at or near the two major proximal chromosome breakpoints (BP1 and BP2) and the distal breakpoint (BP3) in the 15q11-q13 region [80-82]. The typical PWS deletion consists of two classes, Type I and Type II. The larger Type I (TI) deletion involves breakpoint 1 (BP1) which is nearest the centromere while the smaller Type II (TII) deletion involves breakpoint 2 (BP2). Breakpoint 3 (BP3) is located distally in the 15q11-q13 region and is common in both typical deletion subgroups. Butler and colleagues [79] reported that Type I deletions were approximately 6.6 Mb in size and the Type II deletion was 5.3 Mb. There are four recognized genes located between BP1 and BP2 (*i.e.*, GCP5, CYFIP1, NIPA1, NIPA2) and individuals were reported recently with behavioral and autistic findings with a deletion involving only the region between BP1 and BP2 [26, 28]. PWS individuals with the larger typical Type I deletion (involving BP1 and BP3) appear to have more behavioral problems such as obsessive compulsion and self-injury along with lower scores on measures of academic performance than seen in those PWS individuals with the smaller Type II deletion (involving BP2 and BP3) [20]. Atypical deletions involving the 15q11-q13 region that are greater or smaller in size than the typical Type I or Type II deletions have been reported in about 5% of PWS individuals [11, 22].

There are several dozen genes or transcripts identified in the 15q11-q13 region located between the breakpoints BP1 and BP3 in the 15q11-q13 region. About 10 of these genes are imprinted and paternally expressed only [2]. DNA methylation analysis of genes in the region is accurate for diagnosis in about 99% of individuals with PWS. However, this testing will not identify the specific genetic subtype (deletion, maternal disomy or an imprinting defect), where separate testing is required to determine the specific subtype, including fluorescence *in situ* hybridization (FISH), genotyping with informative chromosome 15 DNA markers, Methylation specific-multiplex ligation dependent probe amplification MS-MLPA kits and chromosomal or high resolution microarrays with copy number variant (CNV) and single nucleotide polymorphism (SNP) probes [83, 84] (see Fig. **2**).

The 15q11-q13 region can be divided into four regions; (1) a proximal non-imprinted region between BP1 and BP2 containing the four biparentally expressed genes; (2) a 'PWS paternal-only expressed region' containing five well characterized protein coding genes (MKRN3, MAGEL2, NDN, and the bicistronic SNURF-SNRPN), a cluster of five snoRNA genes (HBII-436, HBII-

13, HBII-438, HBII-85 and HBII-52) and several antisense transcripts including the antisense transcript to UBE3A; (3) an 'AS region' containing the preferentially maternally expressed genes UBE3A and ATP10A and (4) a distal non-imprinted region containing a cluster of three GABA receptor genes, the P gene for oculocutaneous albinism type 2 (OCA2) and the HERC2 gene [2, 11].

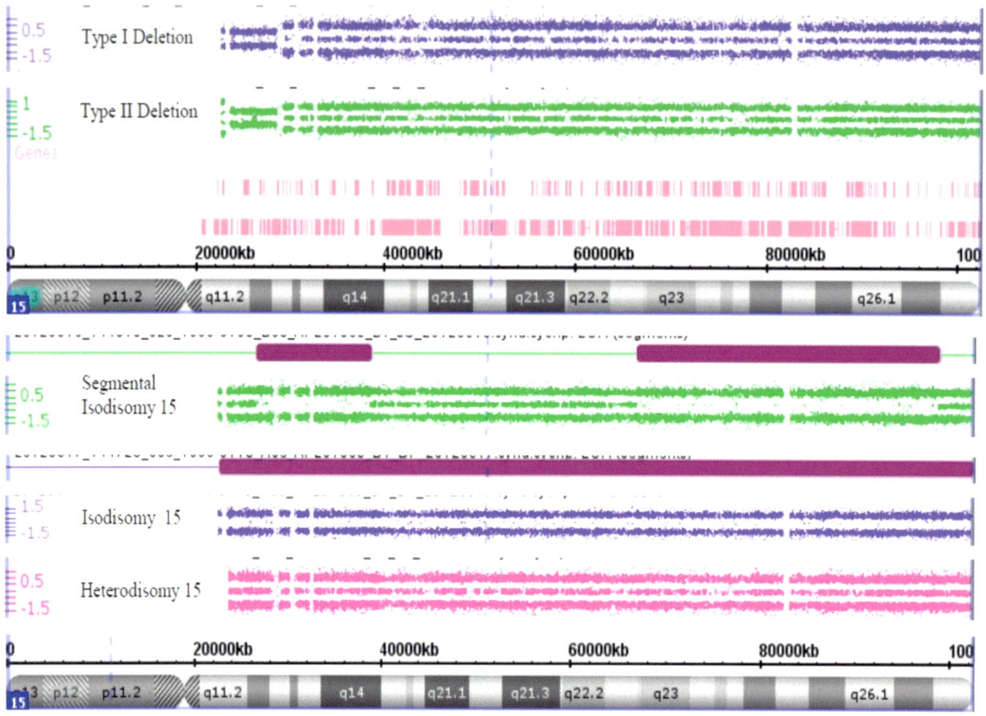

Fig. (2). A: High resolution microarray using CNV and SNP probes to identify the type I and type II 15q11-q13 deletion in PWS. **B:** High resolution microarray using CNV and SNP probes to identify maternal disomy 15 subtype (segmental isodisomy, isodisomy, heterodisomy).

The complex SNRPN (small nuclear ribonucleoprotein N) locus and a second protein coding sequence (SNURF, or SNRPN upstream reading frame) are located in the 15q11-q13 region in as well as multiple copies of C/D box small nucleolar RNAs (snoRNAs) or SNORDs involved in RNA processing encoded within the SNURF-SNRPN transcript [85-87]. Other snoRNAs included in this region are SNORD64, SNORD107, SNORD109A, SNORD115 and SNORD116 previously referred to as HBII-13, HBII-436, HBII-438, HBII-52 and HBII-85. The MKRN3, MAGEL2, NDN and C15orf2 genes are involved in brain development and function. Exons 4-10 of the complex bicistronic SNURF–SNRPN gene encodes for a core spliceosomal protein (SmN) involved in mRNA splicing in the brain, whereas exons 1–3 encode a 71-amino- acid protein enriched in arginine residues [2, 11, 86]. Necdin (NDN) is a paternally expressed gene involved with axonal outgrowth and is overly expressed in the hypothalamus, thalamus and pons. Mice deficient for the necdin gene show delayed migration of the sympathetic neurons, neonatal lethality and respiratory problems. The MAGEL2 gene is paternally expressed in various brain regions including the hypothalamus and appears to play a role in circadian rhythm, brain structure, behavior, and fertility in humans. Recently, MAGEL2 gene mutations have been reported in a subset of individuals with autism [88]. The MKRN3 gene is a member of the makorin (MKRN) RING finger protein gene family which encodes specific proteins (makorins) and abundantly expressed in the central nervous system. The MKRN3 gene has also been reported as a causation of precocious puberty in families without PWS [89]. Two additional potential genes (PWRN1 and PWRN2) were recently identified and located close to the NDN gene. PWRN1 appears as a novel alternative start site for SNURF- SNRPN while PWRN2 is male germ cell-specific and expressed from the haploid genome [10, 11, 86, 87, 90].

The two imprinted genes (UBE3A, ATP10C) are maternally expressed or paternally silenced and located in this chromosome region. The UBE3A gene causes Angelman syndrome. Other genes located in the distal area of the 15q11-q13 region include several gamma aminobutyric acid (GABA) receptor subunits called GABRB3, GABRA5, GABRG3 which appear to be paternally biased with loss of the paternal allele resulting in a reduction of expression of greater than 50% [78]. GABA is a major inhibitory neurotransmitter and alterations associated with hunger, obsessive-compulsions and abnormal visual perception and memory; all present in PWS [1-3]. The P gene for pigment production is also located in the distal end of this chromosome region and deleted in both PWS and AS individuals. Mutations of this gene cause albinism.

GENE/TRANSCRIPT EXPRESSION IN PRADER- WILLI AND ANGELMAN SYNDROMES

To further characterize the genetic findings and identify unique expression patterns in PWS and AS, custom-made cDNA and whole genome-wide expression microarrays were used with RNA isolated from lymphoblastoid cell lines established from three comparison males and seven PWS males with the 15q11-q13 deletion and maternal disomy 15 [66]. More than 47,000 probes were examined and 23,383 were detectable. As expected, imprinted genes from the 15q11-q13 region (*e.g*, SNRPN, NDN, MAGEL2) showed no expression. In addition, there was no difference in expression levels for biallelically expressed genes (*e.g.*, OCA2) within the 15q11-q13 region when comparing maternal disomy 15 cell lines with controls. However, two genes previously identified as maternally expressed (UBE3A, ATP10C) showed a significant increase in expression level in maternal disomy cell lines compared with control or PWS deletion subjects. Several genes/transcripts (*e.g.*, GABRA5, GABRB3) showed increased expression in the PWS maternal disomy cells compared with PWS deletion cells, but less than in control cells, indicating a paternal bias of expression [78].

Significant differences (1.5 fold change with false discovery rate of 20%) were found when placed in functional categories for 323 genes with expression in lymphoblastoid cells from males with PWS relative to comparison males characterized with the Ingenuity Pathway Analysis computer software program. These categories included the nervous system development and function, tissue morphology, hematological system development and function and regulation of gene expression [66]. Genes showing significant changes in expression in the PWS males other than those genes known to be imprinted and paternally expressed included HCRT [hypocretin (orexin)], POMC (pro-opiomelanocortin, OXTR (oxytocin receptor), MC2R (melanocortin receptor) and serotonin receptors 3A, 1B and 2B; all are involved with eating behavior in humans. Other genes of interest with reduced expression in PWS individuals included STAR (a key regulator of steroid synthesis) and SAG (an arrestin family member which desensitizes G- protein-coupled receptors). The impact of the 15q11-q13 genes involved with coding or non-coding protein production on the function of genes located elsewhere in the genome and their role in the causation of PWS will require further investigations.

Obesity with hyperphagia is seen in several rare obesity-related genetic syndromes including Prader–Willi syndrome (PWS) and Alström syndrome (ALMS) [91]. We compared coding and noncoding expression patterns using expression microarrays in adult males with PWS, ALMS, and nonsyndromic obese and nonobese males from RNA isolated from lymphoblastoid cells. We proposed to identify disease-specific molecular patterns and disturbed mechanisms in obesity in these obesity-related genetic syndromes. Over 200 genes were upregulated in ALMS compared with nonobese males, but no genes were found to be upregulated in obese or PWS males and 124 genes were downregulated in ALMS. Only the complex SNRPN locus was disturbed (downregulated) in PWS along with several downregulated small nucleolar RNAs (snoRNAs) in the 15q11-q13 region (SNORD116, SNORD109B, SNORD109A,

SNORD107. Fifty-two miRNAs associated with multiple, overlapping gene expression disturbances were upregulated in ALMS, and four were shared with obese males but not PWS males. Seven passenger strand microRNAs (miRNAs) (miR-93*, miR-373*, miR-29b-2*, miR-30c-1*, miR27a*, miR27b*, and miR-149*) were disturbed in association with six separate downregulated target genes (CD68, FAM102A, MXI1, MYO1D, TP53INP1, and ZRANB1). Cell cycle (*e.g.*, PPP3CA), transcription (*e.g.*, POLE2), and development may be impacted by upregulated genes in ALMS, while downregulated genes were found to be involved with metabolic processes (*e.g.*, FABP3), immune responses (*e.g.*, IL32), and cell signaling (*e.g.*, IL1B*)*. The high number of gene and noncoding RNA disturbances in ALMS contrast with observations in PWS and males with nonsyndromic obesity and may reflect the progressing multiorgan pathology of the ALMS disease process.

ATYPICAL DELETIONS IN PRADER-WILLI SYNDROME: CLINICAL FINDINGS

In 1996, Butler and coworkers [23] reported on a 5-year- old white girl with features of PWS identified during infancy. She was found to have a submicroscopic deletion of the 15q11-q13 region using FISH and DNA marker analysis. The deletion was approximately 100-200 kb in size and showed a *de novo* paternal deletion of the imprinting center and the SNRPN gene region including SNORD116. She developed a milder PWS phenotype with decreased growth, cognitive and behavioral problems than typically seen in PWS. She also had normal pigmentation.

More recently, a female patient with an atypical deletion of the chromosome 15q11-q13 region was reported due to an unbalanced chromosome 15 translocation involving only a paternal deletion of the MKRN3, MAGEL2 and NDN genes [92]. She presented with obesity, mental retardation and a high pain threshold but lacked other features of PWS. Other patients included a male child with several features of PWS reported by Sahoo and colleagues in 2008 [93]. The features included neonatal hypotonia, feeding problems, obesity, and hypogenitalism, but with increased birth weight, a large head and facial appearance not consistent with PWS. Genetic testing showed a paternal deletion of snoRNA SNORD116 and part of the SNORD115 gene cluster. De Smith and colleagues [94] reported a 19-year-old male with hyperphagia and severe obesity, mild intellectual disability and hypogonadism with a 187 kb deletion which included SNORD116. Deletions involving the proximal long arm of chromosome 15 can be smaller or larger than the typical deletion in individuals with features of

PWS as supported by a recent report by Butler and colleagues [24] of an infant girl with a *de novo* interstitial 15q11-q14 deletion, which is larger than the typical deletion. She presented with hypotonia, a poor suck, feeding problems, mild micrognathia, preauricular ear tags, a high arched palate, edematous feet and coarctation of the aorta; several of these features are not typical for PWS but reported previously in individuals with this expanded deletion of chromosome 15.

Clinical reports of atypical deletions (larger and smaller) give further insight into the genes causing PWS. The very small atypical deletions appear to contribute to several PWS features, specifically the paternally expressed snoRNAs including SNORD115 and SNORD116. In addition, mice with the snoRNA equivalent of SNORD116 removed causes hyperphagia and growth deficiency. Furthermore, Kishore and Stamm [95] reported that the paternally expressed snoRNA SNORD115 regulates alternative splicing of the serotonin 5-HT2C receptor leading to an abnormal receptor involved with eating behavior. Therefore, PWS appears not to be caused by a single locus or gene defect, but by a deficiency of a combination of imprinted genes in the 15q11-q13 region including SNURF-SNRPN and the SNORD genes. Deficiencies of other imprinted genes in the region, including MKRN3, MAGEL2 and/or NDN apparently play a role but not sufficient alone to generate the cardinal features recognized in Prader-Willi syndrome.

CLINICAL COMPARISON OF GENETIC SUBTYPES IN PRADER-WILLI SYNDROME

Individuals with PWS who have the typical chromosome 15 deletion are hypopigmented and more homogeneous in their clinical presentation [1, 15]. Those with maternal disomy have fewer typical facial features and are less likely to have certain behavioral findings such a skin picking, unusual skill with jigsaw puzzles, a high pain threshold and articulation problems. PWS individuals with maternal disomy are usually diagnosed later, which may reflect a milder phenotype. PWS individuals with the 15q11-q13 deletion have a greater number of compulsive symptoms than those with maternal disomy 15 [21, 77]. Furthermore, PWS children with the deletion possess relative strengths in standardized visual-spatial tasks such as object assembly [42].

Measures of intellectual function and academic achievement in PWS have shown that those with maternal disomy have significantly higher verbal IQ scores than those with the typical deletion [41]. However, PWS deletion individuals score higher on object-assembly subtests, supporting specific visual perceptual skills as relative strengths in these individuals. Discrimination of shape and motion testing and kinetic form visual performance in the maternal disomy group are generally worse than seen in the deletion individuals. However, the superior visual recognition memory is greater in the maternal disomy individuals [96].

Table 1. Reported clinical differences in genetic subtypes of Prader-Willi syndrome.

Genetic Defect	Characteristics
Typical 15q11-q13 deletion	Hypopigmentation and more homogenous clinical findings including a typical facial appearance; more self-injurious behavior (skin picking); higher pain threshold; greater jigsaw puzzle skills than seen in maternal disomy
Type I deletion	Increased maladaptive and compulsive behavior relative to Type II deletion and maternal disomy; poorer academic performance relative to Type II deletion and maternal disomy
Type II deletion	Better adaptive behavior and social skills relative to Type I or maternal disomy
Maternal disomy 15	Higher verbal IQ scores; greater numeric calculation skills, superior visual memory, poorer object assembly and visual perceptual skills; increased psychosis relative to typical deletion

To further characterize clinical differences in the genetic subtypes in PWS, Butler and colleagues [20] reported differences in psychological, cognitive and behavioral data collected from young adults with PWS representing the longer Type I deletion involving breakpoints BP1 and BP3 *versus* those with the shorter Type II deletion involving BP2 and BP3. PWS individuals with Type I deletions scored significantly worse in self-injurious and maladaptive behavior assessments compared with Type II deletions. Obsessive–compulsive behavior was also more common in Type I deletions. Adaptive behavior scores were generally more abnormal including lower reading and math skills and more compulsions in Type I *versus* Type II deletion individuals. Intellectual ability and academic achievement were lower in individuals with Type I deletion (see Table **1**).

To help explain the genetic deletion subtype differences, Bittel and colleagues [97] studied expression patterns of four genes (GCP5, CYFIP1, NIPA2, NIPA1) located between BP1 and BP2 and correlated the expression with behavioral, maladaptive and intellectual ability scores in the two PWS deletion groups. Three of these genes are implicated in central nervous system development. Correlation studies with gene expression and behavioral and cognitive scores indicated a direct relationship between the quantity of gene message and higher assessment scores. The coefficient of determination indicated the quantity of messenger RNA of the four genes explained 24% to 99% of the variation of the behavioral and academic assessment scores, with NIPA2 having the greatest impact.

In summary, advances in genetic testing have led to a better understanding of the role of genetics and gene disturbances with impact on clinical care in PWS and for other genetic conditions. The publication of Health Supervision for Children with Prader-Willi syndrome published in Pediatrics [98] and the use of interactive computer websites such as the Prader-Willi Syndrome Diagnosis Module in the Medical Home Portal (www. medcalhomeportal.org) can provide useful information to care providers treating individuals with PWS. Other internet sites include the Prader-Willi Syndrome Association (USA) at www. pwsausa.org; Foundation for Prader-Willi Research at www.fpwr.org; Foundation for Prader-Willi Research Canada at www.onesmarcstep.ca and the International Prader-Willi Syndrome Organization at www.ipwso.org. These are excellent online sources of information for the parents and health care providers needed for management and care of individuals at all ages with Prader-Willi syndrome.

ACKNOWLEDGEMENTS

I thank Carla Meister, Tiffany Embry and Rebecca Wall for graphic illustration production and preparation of manuscript including F. John Meaney, Ph.D. for helpful comments and suggestions. I acknowledge the National Institutes of Health grant NICHD HD02528.

CONFLICT OF INTEREST

The author confirms that he has no conflict of interest to declare for this publication.

DISCLOSURE

This is an updated chapter of the earlier review article [by Butler, M. G. entitled "Prader-Willi syndrome: obesity due to genomic imprinting". *Current Genomics*, 2011; 12(3), 204-215] with additional text and references.

ABBREVIATIONS

AS	=	Angelman syndrome
DNA	=	Deoxyribonucleic acid
fMRI	=	Functional magnetic resonance imaging
IQ	=	Intelligence quotient

MS-MLPA = Methylation specific-multiplex ligation dependent probe amplification

PWS = Prader-Willi syndrome

UPD = Uniparental disomy

REFERENCES

[1] Butler MG. Prader-Willi syndrome: Current understanding of cause and diagnosis. Am J Med Genet 1990; 35(3): 319-32.

[2] Butler MG. Prader-Willi syndrome: Obesity due to genomic imprinting. Curr Genomics 2011; 12: 204-15.

[3] Cassidy SB, Schwartz S, Miller JL, Driscoll DJ. Prader-Willi syndrome. Genet Med 2012; 14(1): 10-26.

[4] Butler MG, Thompson T. Prader-Willi syndrome: Clinical and genetic finding. Endocrinology 2000; 10: 3S-16S.

[5] Whittington J, Holland A, Webb T, Butler J, Clarke D, Beer H. Cognitive abilities and genotype in a population-based sample of people with Prader-Willi syndrome J Intellect. Disabil Res 2004; 48(Pt 2): 172-87.

[6] Butler MG, Weaver DD, Meaney FJ. Prader-Willi syndrome: Are there population differences? Clin Genet 1982; 22(5): 292-4.

[7] Butler MG, Lee PDK, Whitman BY. Management of Prader-Willi Syndrome. Springer: New York 2006.

[8] Ohta T, Gray TA, Rogan PK, Buiting K, Gabriel JM, Saitoh S, *et al.* Imprinting-mutation mechanisms in Prader-Willi syndrome. Am J Hum Genet 1999; 64(2): 397-413.

[9] Buiting K, Gross S, Lich C, Gillessen-Kaesbach G, el-Maarri O, Horsthemke B. Epimutations in Prader-Willi and Angelman syndromes: A molecular study of 136 patients with an imprinting defect. Am J Hum Genet 2003; 72(3): 571-7.

[10] Hanel ML, Wevrick R. The role of genomic imprinting in human developmental disorders: Lessons from Prader-Willi syndrome. Clin Genet 2001; 59(3): 156-64.

[11] Bittel DC, Butler MG. Prader-Willi syndrome: Clinical genetics, cytogenetics and molecular biology. Expert Rev Mol Med 2005; 7(14): 1-20.

[12] Nicholls RD, Knoll JH, Butler MG, Karam S, Lalande M. Genetic imprinting suggested by maternal heterodisomy in nondeletion Prader-Willi syndrome. Nature 1989; 342(6247): 281-5.

[13] Williams CA, Driscoll DJ, Dagli AI. Clinical and genetic aspects of Angelman syndrome. Genet Med 2010; 12(7): 385-95.

[14] Butler MG. Genomic imprinting disorders in humans: A mini-review. J Assist Reprod Genet 2009; 26(9-10): 477-86.

[15] Butler MG. Hypopigmentation: A common feature of Prader-Labhart-Willi syndrome. Am. J. Hum Genet 1989; 45(1): 140-6.

[16] Butler MG, Meaney FJ, Palmer CG. Clinical and cytogenetic survey of 39 individuals with Prader-Labhart-Willi syndrome. Am J Med Genet 1986; 23(3): 793-809.

[17] Prader A, Labhart A, Willi H. Ein sydnrom von adipositas, kleinwuchs, kryptorchismus undoligophrenie nach myatonieartigem zustand im neugeborenenalter. Schweizerische Medizinishce Wochenschrift 1956; 6(3): 1260-1.

[18] Ledbetter DH, Riccardi VM, Airhart SD, Strobel RJ, Keenan BS, Crawford JD. Deletions of chromosome 15 as a cause of the Prader-Willi syndrome. N Engl J Med 1981; 304(6): 325-9.

[19] Butler MG, Palmer CG. Parental origin of chromosome 15 deletion in Prader-Willi syndrome. Lancet 1983; 1(8336): 1285-6.

[20] Butler MG, Bittel DC, Kibiryeva N, Talebizadeh Z, Thompson T. Behavioral differences among subjects with Prader-Willi syndrome and type I or type II deletion and maternal disomy. Pediatrics 2004; 113(3 Pt 1): 565-73.

[21] Zarcone J, Napolitano D, Peterson C, Breidbord J, Ferraioli S, Caruso-Anderson M, *et al.* The relationship between compulsive behaviour and academic achievement across the three genetic subtypes of Prader-Willi syndrome. J Intellect Disabil Res 2007; 51(Pt. 6): 478-87.

[22] Kim SJ, Miller JL, Kuipers PJ, German JR, Beaudet AL, Sahoo T, *et al.* Unique and atypical deletions in Prader-Willi syndrome reveal distinct phenotypes. Hum Mol Genet 20 2012; (3): 283-90.

[23] Butler MG, Christian SL, Kubota T, Ledbetter DH. A 5- year-old white girl with Prader- Willi syndrome and a submicroscopic deletion of chromosome 15q11q13. Am J Med Genet 1996; 65(2): 137-41.

[24] Butler MG, Bittel DC, Kibiryeva N, Cooley LD, Yu S. An interstitial 15q11-q14 deletion: Expanded Prader-Willi syndrome phenotype. Am J Med Genet A 2010; 152A(2): 404-8.

[25] Svenstrup K, Moller RS, Christensen J, Budtz-Jorgensen E, Gilling M, Nielsen JE. NIPA1 mutation in complex hereditary spastic paraplegia with epilepsy. European journal of neurology. Eur J Neurol 2011; 18 (9): 1197-9.

[26] Burnside RD, Pasion R, Mikhail FM, Carroll AJ, Robin NH, Youngs EL, *et al.* Microdeletion/microduplication of proximal 15q11.2 between BP1 and BP2: A susceptibility region for neurological dysfunction including developmental and language delay. Hum Genet 130 2011; 130(4): 517-28.

[27] Jerkovich AM, Butler MG. Further phenotypic expansion of 15q11.2 BP1-BP2 microdclction (Burnside-Butler) syndrome. J Pediatr Genet 2014; 3(1): 41-4.

[28] Cox DM, Butler MG. The 15q11.2 BP1-BP2 microdeletion syndrome: A review. Int J Mol Sci 2015; 16(2): 4068-82.

[29] Miller JL, Lynn CH, Driscoll DC, Goldstone AP, Kimonis V, Dykens E, *et al.* Nutritional phases in Prader- Willi syndrome. Am J Med Genet A 2011; 155A(5): 1040-9.

[30] de Lind van Wijngaarden RF, Joosten KF, van den Berg S, Otten BJ, de Jong FH, Sweep CG, *et al.* The relationship between central adrenal insufficiency and sleep-related breathing disorders in children with Prader-Willi syndrome. J Clin Endocrinol Metab 2009; 94(7); 2387-93.

[31] Sedky K, Bennett DS, Pumariega A. Prader-Willi syndrome and obstructive sleep apnea: Cooccurrence in the pediatric population. J Clin Sleep Med 2014; 10(4): 403-409.

[32] Beebe D. Neurobehavioral morbidity associated with disordered breathing during sleep in children: A comprehensive review. SLEEP 2006; 29(9): 1115-34.

[33] Maas AP, Sinnema M, Didden R, Maaskant MA, Smits MG, Schrander-Stumpel CT, *et al.* Sleep disturbances and behavioral problems in adults with PWS. J Intellect Disabil Res 2010; 54: 906-17.

[34] Butler MG, Sturich J, Lee J, Myers SE, Whitman BY, Gold JA, *et al.* Growth standards of infants with Prader-Willi syndrome. Pediatrics 2011; 127 (4): 687-95.

[35] Butler MG. Management of obesity in Prader-Willi syndrome. Nat Clin Pract Endocrinol Metab 2006; 2(11): 592-3.

[36] Goldstone AP, Holland AJ, Hauffa BP, Hokken-Koelega AC, Tauber M. Recommendations for the diagnosis and management of Prader-Willi syndrome. J Clin Endocrinol Metab 2008; 93(11): 4183-97.

[37] Butler MG, J Roberts, J Hayes, X Tan, Manzardo AM. Growth hormone receptor (GHR) gene polymorphism and Prader-Willi syndrome. Am J Med Genet A 2013; 161A (7): 1647-53.

[38] Fox R, Sinatra RB, Mooney MA, Feurer ID, Butler MG. Visual capacity and Prader-Willi syndrome. J Pediatr Ophthalmol Strabismus 1999; 36(6): 331-6.

[39] Diene G, Mimoun E, Feigerlova E, Caula S, Molinas C, Grandjean H, *et al.* French Reference Centre for PWS: Endocrine disorders in children with Prader-Willi syndrome - data from 142 children of the French database. Horm Res Paediatr 2010; 74:121-8.

[40] Irizarry K, Bain J, Butler MG, Ilkayeva O, Newgard C, Muehlbauer M, *et al.* Metabolic profiling in Prader-Willi syndrome and non-syndromic obesity: Sex differences and the role of growth hormone treatment. Clin Endocrinol (oxf) 2015; doi:10.1111/cen.12766. [Epub ahead of print].

[41] Roof E, Stone W, MacLean W, Feurer ID, Thompson T, Butler MG. Intellectual characteristics of Prader-Willi syndrome: comparison of genetic subtypes. J Intellect Disabil Res 2000; 44(Pt 1): 25-30.

[42] Dykens EM. Are jigsaw puzzle skills 'spared' in persons with Prader-Willi syndrome? J Child Psychol Psychiatry 2002; 43(3): 343-52.

[43] Crino A, Schiaffini R, Ciampalini P, Spera S, Beccaria L, Benzi F, *et al.* Hypogonadism and pubertal development in Prader-Willi syndrome. Eur J Pediatr 2003; 162(5): 327-33.

[44] Greenswag LR. Adults with Prader-Willi syndrome: A survey of 232 cases. Dev Med Child Neurol 1987; 29(2): 145-52.

[45] Butler MG, Meaney FJ. Standards for selected anthropometric measurements in Prader-Willi syndrome. Pediatrics 1991; 88(4): 853-60.

[46] Angulo M, Castro-Magana M, Mazur B, Canas JA, Vitollo PM, Sarrantonio M. Growth hormone secretion and effects of growth hormone therapy on growth velocity and weight gain in children with Prader-Willi syndrome. J Pediatr Endocrinol Metab 1996; 9(3): 393-400.

[47] Lindgren AC, Hagenas L, Muller J, Blichfeldt S, Rosenborg M, *et al.* Children with Prader-Willi syndrome affects linear growth and body composition favourably. Acta Paediatr 1998; 87(1): 28-31.

[48] Lee PDK. Endocrine and metabolic aspects of Prader-Willi syndrome. In: Management of Prader- Willi syndrome, 2nd ed Greenswag LR, Alexander, RC, Eds. Springer-Verlag: New York 1995.

[49] Burman P, Ritzen EM, Lindgren AC. Endocrine dysfunction in Prader-Willi syndrome: A review with special reference to GH. Endocr Rev 2001; 22(6): 787-99.

[50] Butler MG, Lee J, Manzardo AM, Gold JA, Kimonis V, Miller JL, *et al.* Growth charts for nongrowth hormone treated Prader-Willi syndrome. Pediatrics 2015; 135: e126-35.

[51] Oto Y, Tanaka Y, Abe Y, Obata K, Tsuchiya T, Yoshino A, *et al.* Exacerbation of BMI after cessation of growth hormone therapy in patients with Prader-Willi syndrome. Am J Med Genet A 2014; 164A(3): 671-5.

[52] Coupaye M, Lorenzini F, Lloret-Linares C, Molinas C, Pinto G, Diene G, *et al.* Growth hormone therapy for children and adolescents with Prader-Willi syndrome is associated with improved body composition and metabolic status in adulthood. J Cli Endocrinol Metab 2013; 98: E328-35.

[53] Butler MG, Smith BK, Lee J, Gibson C, Schmoll C, Moore WV, *et al.* Effects of growth hormone treatment in adults with Prader-Willi syndrome. Growth Horm IGF Res 2013; 23(3): 81-7.

[54] Mogul HR, Lee PD, Whitman BY, Zipf WB, Frey M, Myers S, *et al.* Growth hormone treatment of adults with Prader-Willi syndrome and growth hormone deficiency improves lean body mass, fractional body fat, and serum triiodothyronine without glucose impairment: results from the United States multicenter trial. J Clin Endocrinol Metab 2008; 93(4): 1238-45.

[55] Sode-Carlsen R, Farholt S, Rabben KF, Bollerslev J, Schreiner T, Jurik AG, *et al.* Growth hormone treatment in adults with Prader-Willi syndrome: the Scandinavian study. Endocrine 2012; 41(2): 191-9.

[56] Sode-Carlsen R, Farholt S, Rabben KF, Bollerslev J, Schreiner T, Jurik AG, *et al.* One year of growth hormone treatment in adults with Prader-Willi syndrome improves body composition: Results from a randomized, placebo-controlled Study. J Clin Endocrinol Metab 2010; 95(11): 4943-50.

[57] Sanchez-Ortiga R, Klibanski A, Tritos NA. Effects of recombinant human growth hormone therapy in adults with Prader-Willi syndrome: A meta-analysis. Clin Endocrinol 2012; 77(1): 86-93.

[58] Molitch M, Clemmons D, Malozowski S, Merriam G, Vance M. Evaluation and treatment of adult growth hormone deficiency: An endocrine society clinical practice guideline. J Clin Endocrinol Metab 2011; 96(6): 1587-609.

[59] Stevenson DA, Heinemann J, Angulo M, Butler MG, Loker J, Rupe N, *et al.* Gastric rupture and necrosis in Prader-Willi syndrome. J Pediatr Gastroenterol Nutr 2007; 45(2): 272-4.

[60] Boer H, Holland A, Whittington J, Butler J, Webb T, Clarke D. Psychotic illness in people with Prader Willi syndrome due to chromosome 15 maternal uniparental disomy. Lancet 2002; 359(9301): 135-6.

[61] Vogels A, De Hert M, Descheemaeker MJ, Govers V, Devriendt K, Legius E, *et al.* Psychotic disorders in Prader-Willi syndrome. Am J Med Genet A 2004; 127A(3): 238-43.

[62] Butler JV, Whittington JE, Holland AJ, Boer H, Clarke D, Webb T. Prevalence of and risk factors for, physical ill-health in people with Prader-Willi syndrome: A population-based study. Dev Med Child Neurol 2002; 44(4): 248-255.

[63] Weselake SV, Foulds JL, Couch R, Witmans MB, Rubin D, Haqq AM. Prader-Willi Syndrome, excessive daytime sleepiness and narcoleptic symptoms: A case report. J Med Case Report 2014; 8:127.

[64] Nishino S, Kanbayashi T. Symptomatic narcolepsy, cataplexy and hypersomnia and their implications in the hypothalamic hypocretin/orexin system. Sleep Med 2005; 9:269-310.

[65] Arii J, Kanbayashi T, Tanabe Y, Sawaishi Y, Kimura S, Watanabe A, *et al.* CSF hypocretin 1 levels in childhood narcolepsy and neurologic disorders. Neurology 2004; 63(12): 2440-2.

[66] Bittel DC, Kibiryeva N, Sell SM, Strong TV, Butler MG. Whole genome microarray analysis of gene expression in Prader-Willi syndrome. Am J Med Genet 2007; 143A(5): 430-2.

[67] Butler MG, Haber L, Mernaugh R, Carlson MG, Price R, Feurer ID. Decreased bone mineral density in Prader-Willi syndrome: Comparison with obese subjects. Am J Med Genet 2001; 103: 216-22.

[68] Butler MG, Theodoro MF, Bittel DC, Donnelly JE. Energy expenditure and physical activity in Prader-Willi syndrome: Comparison with obese subjects. Am J Med Genet A 2007; 143(5): 449-59.

[69] Theodoro MF, Talebizadeh Z, Butler MG. Body composition and fatness patterns in Prader- Willi syndrome: Comparison with simple obesity. Obesity (Silver Spring) 2006; 14(10): 1685-90.

[70] Holsen LM, Zarcone JR, Brooks WM, Butler MG, Thompson TI, Ahluwalia JS, *et al.* Neural mechanisms underlying hyperphagia in Prader-Willi syndrome. Obesity (Silver Spring) 2006; 14(6): 1028-37.

[71] Cummings DE, Clement K, Purnell JQ, Vaisse C, Foster KE, Frayo RS, *et al.* Elevated plasma ghrelin levels in Prader-Willi syndrome. Nat Med 2002; 8(7): 643-4.

[72] Del Parigi A, Tschöp M, Heiman ML, Salbe AD, Vozarova B, Sell SM, *et al.* High circulating ghrelin: A potential cause for hyperphagia and obesity in Prader-Willi syndrome. J Clin Endocrinol Metab 2002; 87: 5461-4.

[73] Butler MG, Bittel DC, Talebizadeh Z. Plasma peptide YY and ghrelin levels in infants and children with Prader-Willi syndrome. J Ped Endo & Metab 2004; 17: 1177-84.

[74] Kweh FA, Miller JL, Sulsona CR, Wasserfall C, Atkinson M, Shuster JJ, *et al.* Hyperghrelinemia in Prader-Willi syndrome begins in early infancy long before the onset of hyperphagia. Am J Med Genet A 2015; 167A(1): 69-79.

[75] Whitman BY, Accardo P. Emotional symptoms in Prader-Willi syndrome adolescents. Am J Med Genet 1987; 28(4): 897-905.

[76] Symons FJ, Butler MG, Sanders MD, Feurer ID, Thompson T. Self-injurious behavior and Prader- Willi syndrome: Behavioral forms and body locations. Am J Ment Retard 1999; 104(3): 260-9.

[77] Dykens EM, Leckman JF, Cassidy SB. Obsessions and compulsions in Prader-Willi syndrome. J Child Psychol Psychiatry 1996; 37(8): 995-1002.

[78] Bittel DC, Kibiryeva N, Talebizadeh Z, Butler MG. Microarray analysis of gene/transcript expression in Prader-Willi syndrome: deletion versus UPD. J Med Genet 2003; 40(8): 568-74.

[79] Butler MG, Fischer W, Kibiryeva N, Bittel DC. Array comparative genomic hybridization (aCGH) analysis in Prader-Willi syndrome. Am J Med Genet A 2008; 146(7): 854-60.

[80] Nicholls RD, Knepper JL. Genome organization, function, and imprinting in Prader-Willi and Angelman syndromes. Annu Rev Genomics Hum Genet 2001; 2: 153-75.

[81] Christian SL, Robinson WP, Huang B, Mutirangura A, Line MR, Nakao M, *et al.* Molecular characterization of two proximal deletion breakpoint regions in both Prader-Willi and Angelman syndrome patients. Am J Hum Genet 1995; 57(1): 40-8.

[82] Chai JH, Locke DP, Greally JM, Knoll JH, Ohta T, Dunai J, *et al.* Identification of four highly conserved genes between breakpoint hotspots BP1and BP2 of the Prader-Willi/Angelman syndromes deletion region that have undergone evolutionary transposition mediated by flanking duplicons. Am J Hum Genet 2003; 73(4): 898-925.

[83] Bittel DC, Kibiryeva N, Butler MG. Methylation-specific multiplex ligation-dependent probe amplification (MLPA) analysis of subjects with chromosome 15 abnormalities. Genet Testing 2007; 11(4): 467-75.

[84] Henkhaus RS, Kim SJ, Kimonis VE, Gold JA, Dykens EM, Driscoll DJ, *et al.* Methylation- specific multiplex ligation-dependent probe amplification and identification of deletion genetic subtypes in Prader-Willi syndrome. Genet Test Mol Biomarkers 2012; 16 (3): 178-86.

[85] Glenn CC, Driscoll DJ, Yang TP, Nicholls RD. Genomic imprinting: Potential function and mechanisms revealed by the Prader-Willi and Angelman syndromes. Mol Hum Reprod 1997; 3(4): 321-32.

[86] Buiting K, Nazlican H, Galetzka D, Wawrzik M, Gross S, Horsthemke B. C15orf2 and a novel noncoding transcript from the Prader-Willi/Angelman syndrome region show monoallelic expression in fetal brain. Genomics 2007; 89(5): 588-95.

[87] Wawrzik M, Spiess AN, Herrmann R, Buiting K, Horsthemke B. Expression of SNURF- SNRPN upstream transcripts and epigenetic regulatory genes during human spermatogenesis. Eur J Hum Genet 2009; 17(11): 1463-70.

[88] Schaaf CP, Gonzalez-Garay ML, Xia F, Potocki L, Gripp KW, Zhang B, *et al.* Truncating mutations of MAGEL2 cause Prader-Willi phenotypes and autism. Nat Genet 45 2013; (11): 1405-8.

[89] Macedo DB, Abreu AP, Reis AC, Montenegro LR, Dauber A, Beneduzzi D, *et al.* Central precocious puberty that appears to be sporadic caused by paternally inherited mutations in the imprinted gene makorin ring finger 3. J Clin Endocrinol Metab 2014; 99(6): E1097-103.

[90] Miller NL, Wevrick R, Mellon PL. Necdin. a Prader-Willi syndrome candidate gene, regulates gonadotropin-releasing hormone neurons during development. Hum Mol Genet 2009; 18(2): 248-60.

[91] Butler MG, Wang K, Marshall JD, Naggert JK, Rethmeyer JA, Gunewardena S, *et al.* Coding and noncoding expression patterns associated with rare obesity-related disorders: Prader-Willi and Alstrom syndromes. Adv Genomics Genet 2015; 5:53-75.

[92] Kanber D, Giltay J, Wieczorek D, Zogel C, Hochstenbach R, Caliebe A, *et al.* A paternal deletion of MKRN3, MAGEL2 and NDN does not result in Prader- Willi syndrome. Eur J Hum Genet 2009; 17(5): 582-90.

[93] Sahoo T, del Gaudio D, German JR, Shinawi M, Peters SU, Person RE, *et al.* Prader-Willi phenotype caused by paternal deficiency for the HBII- 85 C/D box small nucleolar RNA cluster. Nat Genet 2008; 40(6): 719-21.

[94] de Smith AJ, Purmann C, Walters RG, Ellis RJ, Holder SE, Van Haelst, *et al.* A deletion of the HBII- 85 class of small nucleolar RNAs(snoRNAs) is associated with hyperphagia, obesity and hypogonadism. Hum Mol Genet 2009; 18(17): 3257-65.

[95] Kishore S, Stamm S. The snoRNA HBII-52 regulates alternative splicing of the serotonin receptor 2C. Science 2006; 311(5758): 230-2.

[96] Fox R, Yang G, Feurer I, Butler MG, Thompson T. Kinetic form discrimination in Prader- Willi syndrome. J Intell Disabil Res 2001; 45: 317-25.

[97] Bittel DC, Kibiryeva N, Butler MG. Expression of 4 genes between chromosome 15 breakpoints 1 and 2 and behavioral outcomes in Prader-Willi syndrome. Pediatrics 2006; 118 (4): e1276-83.

[98] McCandless SE. Clinical report-health supervision for children with Prader-Willi syndrome. Pediatrics 2011; 127(1): 195-204.

CHAPTER 2

GJB2 Gene Mutations in Syndromic Skin Diseases with Sensorineural Hearing Loss

Sandra Iossa[1,2], Elio Marciano[3] and Annamaria Franzé[1,3,*]

[1]CEINGE-Biotecnologie Avanzate, Naples, Italy; [2]Istituto di Genetica e Biofisica "A. Buzzati Traverso", C.N.R., Naples, Italy and [3]Unità di Audiologia, Dipartimento di Neuroscienze, Scienze Riproduttive e Odontostomatologiche, Università di Napoli "Federico II", Naples, Italy

Abstract: Hearing loss is a very diffuse disease in the population and many cases are associated to genetic causes. Most of the identified mutations have been identified in the *GJB2* gene encoding connexin 26, a transmembrane protein that is a constituent of fundamental structures (gap junctions channels) involved in molecules (ions and small metabolites) exchanges between cells. It has been demonstrated that *GJB2* mutations are causative for different forms of hearing impairment, non-syndromic forms: recessive (DFNB1) and sometimes dominant (DFNA3) as well as syndromic dominant forms with hearing impairment associated to several types of epidermis problems. Connexin 26, in fact, has a relevant role in inner ear functioning but in the skin too, where connexins 26 has been demonstrated to participate in regulation of growth and differentiation of skin. As well as for many cases of hearing loss, skin phenotypes described for connexin 26 mutations are highly variable. Reasons for this high phenotypic variability are actually not clear. This review provides an overview of recent findings concerning pathogenesis of syndromic deafness imputable to *GJB2* mutations with an emphasis on relevant clinical genotype-phenotype correlations. Effects of some mutations on channels function and the relevant role of the hemichannel are discussed.

Keywords: BPS, Calcium Uptake, Cellular Death, Cellular Trafficking, Connexin 26, Deafness, Dominant Mutation, EKV, Epidermis, Gap Junction, Genotype-Phenotype Correlation, *GJB2*, Hearing Loss, Hemichannel, HID, Keratoderma, KID, Phenotypic Variability, PPK, Skin, Vohwinkel Syndrome.

1. INTRODUCTION

Hearing impairment is a very complex pathology and has many and serious consequences for people. It is very diffuse in the population and in many cases it is a congenital disorder affecting 0.1-02% of new-borns [1]. Hearing impairment comprises a broad spectrum of clinical presentation and frequently the cause

***Correspondence author Annamaria Franzè:** Unità di Audiologia, Dipartimento di Neuroscienze, Scienze Riproduttive e Odontostomatologiche, Università degli Studi di Napoli "Federico II", Via Pansini 5, 80131 Napoli, Italy; Tel: 390813737935/390817462990; Fax: 390813737808; E-mail: annamaria.franze@unina.it cannot be identified. It can be subdivided in syndromic and non-syndromic form and, actually, are known at least about 400 syndromic forms. About 60% of the sensorineural forms are ascribed to genetic causes. They show several pattern of inheritance and can be ascribed to more than one hundred genes (http://hereditaryhearingloss.org/). In several cases, in non-syndromic recessive and dominant forms as well as syndromic forms, the pathology can be ascribed to mutations in gap junction protein beta genes and prevalently to the *GJB2* gene (gap junction protein, beta 2) encoding the protein connexin 26 (Cx26).

At the moment, in humans, twenty one genes encoding connexins have been identified [2]. Connexins are tetra-spanning transmembrane proteins sharing the same protein topology. Connexins synthesized in the endoplasmic reticulum (ER) are post-transcriptionally oligomerized in the ER/ Golgi or trans-Golgi network with five other connexin molecules to form a connexon or hemichannels (HCs) (Fig. **1A-B**) [3]. HCs are then trafficked to the plasmatic membrane by vesicular carriers migrating along microtubules. HCs from adjacent cells fuse head-to-head through disulphide bonding to form intercellular channels (gap junctions) with a pore diameter > 1 nm [4] and large gap junctions (GJs) plaques can be formed by the aggregation of several of these channels (Fig. **1C**). These intercellular channels permit the diffusion of several molecules such as ions and small metabolites between cytoplasm of adjacent cells. Plasma membrane hemichannels instead connect the intracellular space with the extracellular milieu. They are generally in a closed physiological state but, in response to some stimulus can function as a direct transmembrane communication pathway, enabling the permeation of ions, but also of several molecules important for intercellular communication and the regulation of inflammatory processes [5]. Aberrant opening of plasma connexons usually leads to cell death and consequently to many diseases [6].

Connexins form a large family of structurally similar gap-junction proteins. Connexin are expressed in several tissues of the body and in several cells are present different types of connexons. So, different connexins can combine in several ways giving origin to different type of GJs and HCs channels with different properties. Hemichannels can be formed from a single species (homomeric) or different Cxs (heteromeric) connexins and complete gap junctions, can be made by analogous (homotypic channels) or diverse homomeric connexons (heterotypic channels), or two heteromeric Cx (heteromeric channels) [7].

Different connexins mutations have been described as causative for human pathologies [8]. Actually it is known that mutation in *GJB2*, *GJE1*, *GJB6*, *GJB4*, *GJB3*, *GJB1*and *GJA1* genes codifying connexins 26, 29, 30, 30.3, 31, 32, and 43 are involved in hearing impairment and some of them (Cx26, Cx30, Cx30.3, Cx31 and Cx43) have been implicated in syndromic form of hearing impairment too affecting internal ear and skin [9].

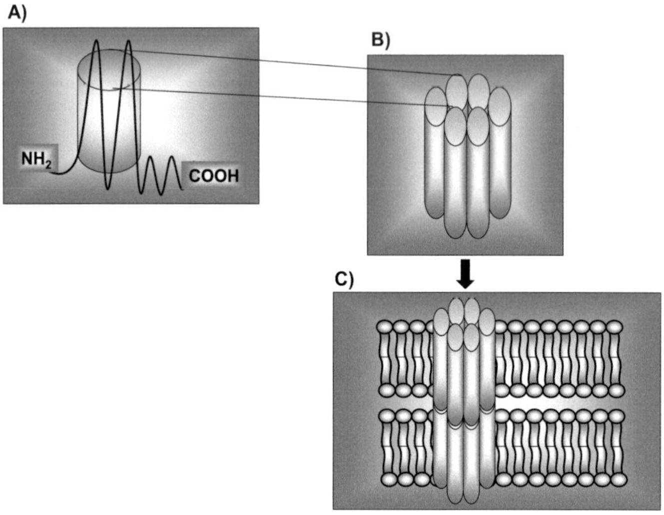

Fig. (1). Graphic structure. Connexins **(A)**, hemichannels formed by six connexins **(B)** and gap junctions formed by two docked hemichannels **(C)**.

A great portion of hereditary forms of sensorineural hearing impairment, non-syndromic recessive and dominant as well as syndromic forms, have been ascribed to mutations in the gene *GJB2* (Cx26) (OMIM 121011) (http://hereditaryhearingloss.org/). In particular, over 50% of congenital recessive non-syndromic hearing loss can be attributed to *GJB2* gene mutations in Caucasian and European populations [10-12]. Mutations in this gene are also implicated in vestibular disorders [13].

2. *GJB2* (CX26) STRUCTURE

The GJB2 gene structure is very simple (Fig. **2**). It is placed on chromosome 13q12 and contains only two exons: an untranslated exon (exon1) separated by an intron of 3179 bp from the coding exon (exon 2).This gene codify the Cx26 protein containing four transmembrane domains (TM1-TM4), two extracellular loops (EC1 and EC2), a cytoplasmic loop (IC2) and cytoplasmic amino terminal (IC1) and carboxy-terminal (IC3) domains [2, 14] (Fig. **3**). The domains IC2 and IC3 are specific of each connexin, while the TM1-TM4 domains and the EC1-EC2 domains are extremely preserved among the complete connexin family. The extracellular domains are relevant to obtain functional intercellular channel while the cytoplasmic portions (IC1-IC2) of the protein are important to obtain the correct channel [15] being involved in the selective intercellular passage of small molecules and control of voltage dependent gating. Recently it has been demonstrated that the N-terminus has a critical role in controlling Cx oligomerization [16] confirming its relevance in formation and proper functioning of GJ.

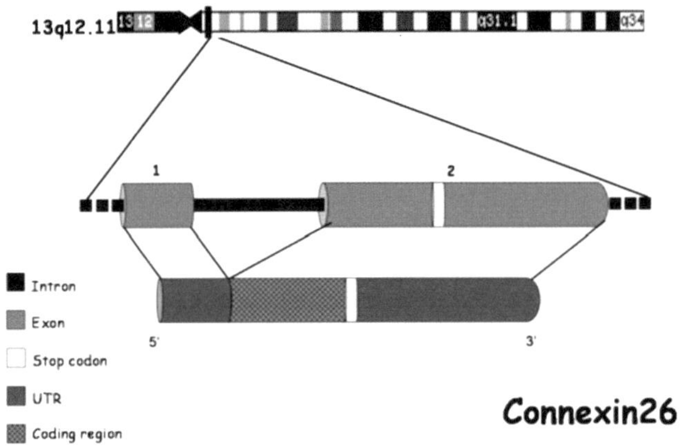

Fig. (2). Graph of *GJB2* molecular structure. *GJB2* gene located on 13q11 chromosome contains only two exons. The exon 1 is a non coding exon while the exon 2 is coding.

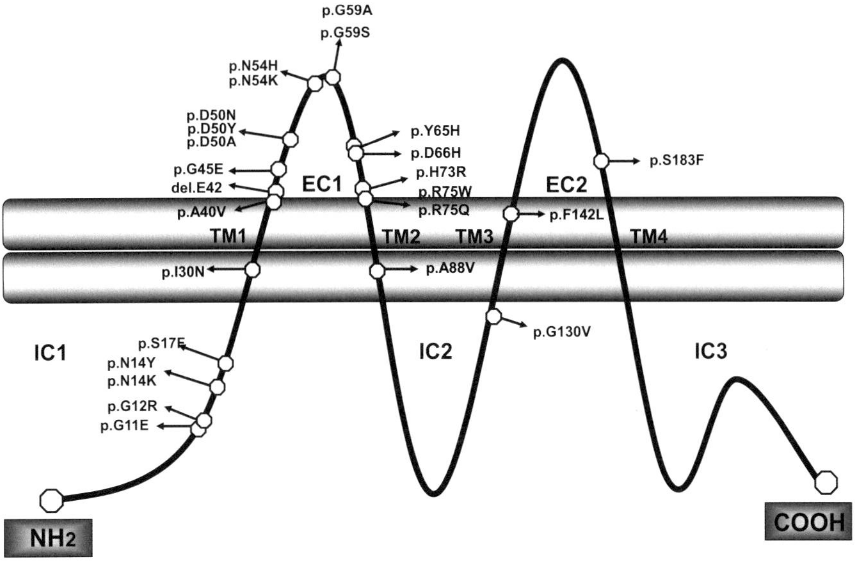

Fig. (3). Connexin 26 structure and known syndromic mutations. Cx26 is formed by four transmembrane domain: TM1-TM4, two extracellular loops: EC1 and EC2, and three intracellular domains: IC1 (NH2-terminal), IC2, and IC3 (COOH-terminal). Along the structure are marked the actually identified mutations related to syndromic forms of hearing impairment.

3. CONNEXIN EXPRESSION IN COCHLEA AND EPIDERMIS

Connexins are expressed in several tissues, but for Cx26 and Cx30 the more relevant role has been demonstrated in inner ear and skin. It has been observed that these two proteins can co-localize and can form heteromeric channels. Cx26 and Cx30 are co-expressed in some parts of the inner ear: in the lateral cochlear wall and also in the organ of Corti, in particular, their expression is present in basolateral and apical regions of the supporting cells but not in hair cells. Their expression has been identified in the spiral ganglion (SG) cells too [17]. At the moment the exact role of the connexins in inner ear functioning is not completely clear. It has been hypothesized that they are important for endolymphatic K+ ions recycling and recently it has been provided the first *in vivo* evidence for active K+ transport through fibrocytes elucidating the mechanism of its contribution to the formation of the endocochlear potential [18]. Moreover, other recent studies have determined other important role for gap junctions in the inner ear. In particular, it has been demonstrated an active role for hemichannels in the release of second messengers and other paracrine messengers fundamental for intercellular exchange in normal ear [19]. Connexin have been divided in two groups (alpha and beta) depending on their structural conformation. In the skin has been identified the expression of several connexins showing sometimes a partially overlapping pattern. Alpha connexins are expressed essentially in non-differentiated basal keratinocyte, while beta connexins are principally present in differentiated keratinocytes. In human skin are present the connexins Cx26, Cx30, Cx30.3, Cx31, Cx31.1, Cx37 and Cx43 [20]. Actually, about Cxs expression in human epidermis, there are no many detailed studies. Some information is present only for Cx43 and Cx26. Cx43 is expressed in interfollicular epidermis, particularly in the spinous and granular cell layers, in sebaceous glands and in hair follicles and plays a relevant role in epidermis development [21]. Cx26 expression is particularly relevant in the skin of palms and soles; its expression has been identified in hair follicles, and in eccrine sweat glands and ducts [22]. Cx26 participates in keratinocyte proliferation control and differentiation and it is expressed during wound healing, psoriasis, and epidermis hyperplasia excited by tumor promoters. In hyperplastic proliferating skin, it has been shown the co-expression of Cx26 and Cx43, but, because Cx26 and Cx43 together cannot realize functional channel, it is possible that their co-expression induces alteration in keratinocytes stimulating proliferation [23, 24].

4. SYNDROMIC FORMS IMPUTABLE TO *GJB2* MUTATIONS

Several GJB2 mutations have been identified in subjects affected with dominant syndromic form of hearing impairment showing epidermis alterations together with inner ear problems. Actually it has been identified only one case that shows a probably recessive mode of inheritance [25]. All the mutations actually identified in the *GJB2* gene are shown in Table **1**. In many cases it is very difficult to realize a clear genotype-phenotype correlation because phenotypic characteristics are highly variable and heterogeneous both for hearing impairment and epidermis symptoms even among subjects carrying identical mutation [26-28]; moreover, in some cases affected subjects show only limited features (Table **2**). More relevant syndromes attributable to Cx26 mutations are: keratitis-ichthyosis-deafness (KID) Syndrome, Hystrix-like Icthyosis Deafness Syndrome (HID), Palmoplantar Keratoderma with Deafness Syndrome, Vohwinkel Syndrome and Bart-Pumphrey Syndrome.

Table 1. Syndromic forms imputable to GJB2 mutations.

Syndrome	Mutations	OMIM code
Keratitis-ichthyosis- deafness (KID)	G11E, G12R, N14K, N14Y, S17F, I30N, A40V, G45E, D50A, D50N, D50Y, A88V	148210
Ichthyosis, hystrix-like-deafness (HID)	D50N, D50Y	602540
Palmoplantar keratoderma-deafness (PPK)	Delta E42, N54H, G59A, H73R, R75Q, R75W, G130V, S183F	148350
Vohwinkel	G59S, Y65H, D66H, G130V	124500
Burt-Pumphrey	N54K, G59S	149200
Unususal mucocutaneous-deafness	F142L	-

The KID syndrome is a rare congenital condition; its main symptoms are represented by skin defects and sensorineural bilateral hearing loss of variable degree. Frequently are present alterations of the surface of the corneas (Keratitis) with photophobia and neovascularization, which result in gradual weakening of visual perception and sometimes in blindness. Epidermal lesions (thickened plaques of skin) are not specific of particular areas of the body and show significant ichthyosis with increased susceptibility to mucocutaneous infection sometimes fatal in the neonatal period [29, 30]. Several other symptoms (alopecia, defective teeth and decreased sweating, dystrophic nails, follicular occlusion, PEODDN, oral mucosal papules and Dandy-Walker malformations) [31, 32] can be present suggesting that the acronym of KID does not include all the features described with the syndrome. In some cases it has been observed an increasing risk to develop cell carcinoma of the skin [33, 34].

Table 2. Major phenotypic characteristics and proteic domains altered by *GJB2* mutations.

Mutations	Syndrome	Phenotypes Ear	 Skin	Protein domain
G11E	KID	Profound SNHL	Hyperkeratosis, ocular problems, alopecia	IC1
G12R	KID	Mild SNHL	Limited hyperkeratosis. Mild ocular problems	IC1
N14K	KID/EKV	Severe SNHL	EKV-like: hypotrichosis, nail dystrophy, mucositis and skin lesions	IC1
N14Y	KID	Profound SNHL	Hyperkeratosis palms and soles, impetiginous plaques on neck, axilla, perianal areas and occipital area, ocular problems	IC1
S17F	KID	SNHL	Visual damage and in one circumstance carcinoma of the tongue; sometimes trichothiodystrophy-like hair abnormalities	IC1
I30N	KID	Profound SNHL	Skin necrosis	TM1
A40V	KID	Profound SNHL	Mild palmoplantar keratoderma, follicular hyperkeratosis, occlusion triad.	TM1/EC1 boundary
Delta E42	PPK	Profound SNHL	Diffuse PPK	TM1/EC1 boundary
G45E	KID	Profound SNHL; inner ear defects.	Severe skin lesion infections and septicaemia:fatal form. In japanese population is not fatal: it is responsible only for a recessive non syndromic form of SNHL.	TM1/EC1 boundary
D50A	KID	Profound SNHL	Hyperkeratosis, corneal abnormalities	EC1
D50N	KID/HID	Profound SNHL and sometimes conductive HL	Photophobia, keratitis, and erythrokeratoderma; sometimes in association with follicular occlusion triad.	EC1
D50Y	KID	Profound SNHL	Photophobia, keratitis, and erythrokeratoderma;	EC1
N54H	PPK	Profound SNHL	PPK and knuckle pads	EC1
N54K	BPS	Profound SNHL	PPK, prominent knuckle pads, and leukonychia	EC1
G59A	PPK	H.F. SNHL	PPK	EC1

Table 2: contd…

G59S	Atypical BPS or Atypical Vohwinkel	Severe SNHL Profound SNHL	Knuckle pads, PPK/ ichthyosis, Massive mutilating keratoderma, proneness to skin cancer	EC1
Y65H	Vohwinkel	Moderate SNHL	Mutilating PPK	EC1
D66H	Vohwinkel	Moderate to severe SNHL	Mutilating PPK	EC1
H73R	PPK	Severe progressive SNHL	Focal PPK similar to Vohwinkel with large intrafamilial variability	EC1
R75W	PPK	Severe to profound SNHL	Variable: generally PPK and some times knuckle pads	EC1/TM2 boundary
R75Q	PPK	Severe to profound SNHL	PPK (not ever present)	EC1/TM2 boundary
A88V	KID	Severe to profound SNHL	Severe skin lesion infections and septicaemia:fatal form.	TM2
G130V	PPK or Vohwinkel	Severe to profound SNHL	Mild PPK PPK and skin constrictions	IC2
F142L	Unususal mucocutaneo us- deafness	Severe to profound SNHL	Psoriasiform mucocutaneous involvement, inflammation of mucous membranes	TM3
S183F	PPK	H. F. SNHL	Focal PPK and some times knuckle pads	EC2

EKV: erythrokeratodermia variabilis; H. F.: high frequency; L. F.: low frequency.

The HID syndrome present many characteristics in common with the KID syndrome but with a dissimilar spreading of hyperkeratosis. It is now supposed that KID and HID might denote the same form of syndromic hearing impairment with a heterogeneous phenotype and that considering the similarity between other syndromic forms of hearing impairment it is possible to divide the syndromic forms in two clusters: the KID-HID cluster and the VS-PPK-BPS cluster [35-37]. The PPK associates to deafness (mild to profound SNHL) conspicuous thickening of the skin principally to the skin on the palms of the hands and the soles of the feet. In the PPK group are present clinically heterogeneous forms of genodermatoses. PPK forms have both recessive as well as dominant type of transmission and some cases of X-linked transmission have been described too. Keratodermas can be clinically classified into diffuse, focal, and punctate [38].

The Vohwinkel syndrome, a rare pathology, generally associates epidermis problems with hearing impairment. Hearing impairment is generally mild or moderate. Skin problems starts in infancy or early childhood with thickened skin present principally on the palms of the hands and soles of the feet (PPK). Hyperkeratosis is present on the knuckles, on the tops of the fingers and toes or on the knees. Skin symptoms are very relevant, in fact, in some cases it is present auto amputation of the digits [20]. The classical form is transmitted in an autosomal pattern and it has been associated to *GJB2* mutation, but a variant form, indicated as loricrin keratoderma, is associated to mutations in the (*LOR*) gene codifying the protein loricrin. Recently, it has been reported a mutilating and focal palmoplantar keratoderma in two siblings with congenital hypotrichosis with probably autosomal recessive inheritance that appears to be a new variant of Vohwinkel syndrome [25].

Bart-Pumphrey syndrome is a very rare pathology starting generally in childhood. In this syndrome are present hearing impairment, PPK, knuckle pads, and white discoloration of the nails (leukonychia) that can also be thick and crumbly [39]. The hearing loss associated with Bart-Pumphrey syndrome ranges from moderate to profound and is typically present from birth.

5. FUNCTIONAL STUDIES

The function of connexins in ear and skin is not completely clear, so to try to have more specific information about their role, several functional studies have been realized about some mutations causing hearing impairment and skin problems. These studies have been realized principally at a cellular level, using either transfected mammalian cells or Xenopus oocyte and recently in bacteria too [40-42]. These studies give more information about the role, assembly and connexin trafficking (Tables **3-4**). In particular, in an interesting study with HeLa cells concerning three Cx26 mutants associated with hearing loss and palmoplantar keratoderma (p.G59A, p.R75Q, and p.R75W) it has been showed that the mutants interact physically with the wild type Cx26 and that the mutants inhibits the transfer of calcein [43]. These results imply a dominant inhibition of the mutated forms on the wild type protein. In another study it has also been demonstrated a trans-dominant inhibition of two Cx26 mutated forms (H73R, S183F) on wild type Cx43 [24].

Moreover, because in the last year is becoming clear a relevant role for hemichannels more than for gap junctions, many studies have been realized about properties of hemichannels (Tables **3** and **4**). Studies, in transgenic mice and transfected primary human keratinocytes, confirm some previous results, obtained in other systems, showing elevated current in some mutated hemichannels [44, 45]. The importance of hemichannel activity in PPK has been, very recently, confirmed in the study of Shuja *et al.* [24] highlighting the importance of Cx43 too in epidermis disease. Moreover, Meigh *et al.* [46] studying the mutant A88V have observed a link between CO_2 sensitivity of Cx26 and skin disease and Levit *et al.* [47] have demonstrated that mefloquine can reduce current membrane alteration in G45E mutants in mouse keratinocytes.

Table 3. Functional activity altered by *GJB2* mutations causative for KID

Mutation	Trafficking	Hemichannels	Functional GJ channels	Observations	Refs.
G11E	Altered	Increased activity	Absent	Increase of calcium uptake, increase of cellular death by necrosis	[15]
G12R	OK	Increased activity	Absent	Increased cellular death. Hyperactive hemichannels when co-expressede with wild type Cx43	[16, 58]
N14K	OK	Increased activity	Present, but with loss of voltage gating	Increased cellular death	[58]
N14Y	-	-	Absent	Hyperactive hemichannels when co-expressed with wild type Cx43	[16, 58]
S17F	OK	Complete loss of activity	Absent	No increased cellular lethality. Hyperactive hemichannels when co-expressede with wild type Cx43	[16, 58]
A40V	OK	Increased activity	Present	Increased cellular death Impaired inhibition by pH and Zn^{2+}	[50, 59]
G45E	OK	Excessive entry of Ca^{2+}	Present	Increased cell death	[44, 59, 60]
D50A	-	Increased activity	-	Increase membrane current flow and increase of cellular death in low cellular calcium solutions	[45]
D50N	-	Altered activity	-	Increase of calcium uptake, increase of cellular death by necrosis and apoptosis	[15, 50, 58]
D50Y	-	Altered activity	-	Increase of calcium uptake, increase of cellular death by necrosis and apoptosis	[61]
A88V	-	Increased activity	-	Increase membrane current flow and increase of cellular death in low cellular calcium solutions	[45]

6. DISCUSSION

Connexins are involved in numerous human biological processes, comprising the regulation of cell growth, differentiation and development. Many effects are determined by metabolic and electrical intercellular coupling. In the last years, it has also been recognized a relevant role for paracrine release of signaling molecules [5].

Several types of connexins are present in many tissues and in several cases it is present a compensatory mechanisms to bypass connexin malfunctioning or absence. However, in some specific organs specific mutations in gap junctions genes can produce serious and devastating diseases. At the moment, it is not clear the precise explanation of the effect of mutations in connexins only in few specific organs. In the last years, many studies have highlighted the importance of Cx26 in the ear and skin. It have been identified many mutations in the *GJB2* gene involved in hearing impairment and skin problems in both non syndromic as well as syndromic forms transmitted with recessive or dominant pattern [6, 8]. The syndromic forms attributable to *GJB2* mutations (Table **1**) associate hearing loss with skin problems and are, generally, transmitted in a dominant manner. The dominant transmission implies in many cases a gain of function that could explain the fact that the altered function of the mutated connexin cannot be overcome by other connexins expressed in the same tissue as instead results in epidermis with homozygous recessive mutations. In skin, in fact, homozygous recessive mutations, that in the ear have a very deleterious effect, don't alter its function and development. Actually, the precise processes involved in diseases associated with *GJB2* mutations continue to be unclear.

By analysing many syndromic cases it is clear that a well-defined genotype-phenotype correlation is very difficult, in fact, notwithstanding the identified mutations regards the same gene (*GJB2*), it is reported a large phenotypic variability with a large variety of clinical signs and rates of progression even in close amino acids (Table **2**). A varying expressivity has been observed even for different mutations in the same amino acid (p.N14K- p.N14Y, p.G59A- p.G59S, p.R75W- p.R75Q) (Table **2**) giving origin to cases with clinical characteristics attributable to different syndromic forms. One suggestion to justify the phenotypic differences observed in these cases may be that distinct substitutions cause different conformational changes to the protein, each with unique consequences for its behaviour acting in some cases on channel formation and in other cases on channels activity. Two interesting situations showing very different phenotype for the same mutations are that observed for G45E and G130V mutations. The first shows the influence of a genetic background on expressivity of a phenotype. In fact, p.G45E in the Japanese patients is causative of a recessive non-syndromic form of hearing impairment, while in other populations it is causative of a fatal form of KID [48] (Table **2**). The second mutation (p.G130V) has been described until now only two times in two different cases: in the first case it has been described as causative of Vohwinkel while in the second case it has been described as causative of PPK (Tables **1** and **2**) [26, 49]. This difference in the observed phenotype for the same mutation reinforces the idea that many described syndromic forms attributable to *GJB2* mutations could have reference to a unique syndromic form with variable phenotype.

Table 4. Functional activity altered by *GJB2* mutations causative for PPK and Vohwinkel mutations.

Mutation	Trafficking	Hemichannels	Functional GJ channels	References
Delta E42	OK	-	Absent	[62]
G59A	OK	Heteromeric: inhibition of calcein transfer	Present	[63]
Y65H	Altered	-	Absent	[64]
D66H	Altered	-	Absent	[65]
H73R	Altered	Absent	Absent	[58, 65]
R75W	OK	Heteromeric: inhibition of calcein transfer	Present	[43, 63]
R75Q	OK	Heteromeric: inhibition of calcein transfer	Present	[43, 64]

In the last years, it have been realized many functional assays in different expression systems to try to understand exact role of connexins in the cells and to have some information useful to explain phenotypic differences observed in mutated connexins. Main part of studies has been realized for KID and PPK/Vohwinkel causative mutations and major results are reported in Tables **3** and in Table **4**. These studies have indicated a relevant role for hemichannels showing an aberrant function for many mutations (Tables **3** and **4**) [9] that may give origin to cell death, due to impaired flux of ions and small metabolites. Studies with the p.G11E and p.D50N mutations have shown rising quantitative of free cytoplasmatic calcium contents and subsequent cell death by necrosis suggesting a relevant role for Cx26 in the control of calcium currents. However, a recent study [50] of A40V, G45E and D50N mutations shows that although each of these mutant hemichannels exhibits impaired inhibition by Ca^{2+}, there is a poor correlation between the degree of impaired inhibition and the severity of the disease phenotype. Thus, additional aberrant hemichannels properties are probably contributing to tissue dysfunction; in fact, A40V mutant shows impaired inhibition by two new possible contributing factors, pH and Zn^{2-}.

Other studies have suggested that some aberrant functions can be influenced by the specific proteic domain altered. It can be observed that almost all KID causative mutations are present in the first 50 amino acid (IC1-TM1-EC1 domains) (Table **2**) while mutations associated to PPK-deafness syndrome are present in the EC1 domain. In this case the only exception is represented by p.G130V that is situated in the IC2 domain bordering TM3 domain [26]. Some studies on mutations situated in EC1 domain suggested that modifications in this domain could impair proper docking of connexons. Also mutations in the IC1 domains are frequently discovered in epidermis pathologies (Table **2**). Among the identified mutations, a relevant role has been assumed for the glycine at position 11 or 12 in maintaining the flexibility of the NT and enabling the gating of the channel; in fact, the substitution of this glycine by a polar residue has been associated with skin disease for several connexins (Cx26 p.G12R in relation to KID, Cx31 p.G12R, p.G12D Cx30.3, p.G12D in EKV, and Cx30 p.G11R in Clouston syndrome) [32, 51-54].

Also some studies [50, 55, 56] on A40V mutation confirm that the effect on functions can depends by the specific domain altered. In fact, these studies suggest that the A40V mutation produces a sufficient perturbation in the structure of the Cx26 hemichannel that affects the properties of the residue that the crystal structure has indicated as part of a hydrophobic core that can function to stabilize subunit structure.

It is clear that many studies have again to be done to clarify many aspects regarding role of connexins in inner ear and skin and the phenotypic differences observed in syndromes associated to *GJB2* mutations. Being observed that Cx26 mutations can alter functions of other Cxs, great attention must be due to interactions of Cx26 with other connexins and other potential partners in different tissues [16, 24, 57].

CONFLICT OF INTEREST

The author confirms that he has no conflict of interest to declare for this publication.

ACKNOWLEDGMENTS

Declared none.

DISCLOSURE

This is an updated Chapter of the earlier review article [by Iossa *et al.* "GJB2 mutations in syndromic skin disorders with sensorineural hearing loss" (review) *Curr Genomics* 2011; 12: 475-485], with additional text and references.

REFERENCES

[1] Pisacane A, Auletta G, Toscano F, *et al.* Feasibility and effectiveness of a population-based newborn hearing screening in an economically deprived region of Italy. Int J Pediatr Otorhinolaryngol 2013; 77(3); 329-33.
[2] Nielsen MS, Axelsen LN, Sorgen PL, *et al.* Gap junctions. Compr Physiol 2012; 2(3): 1981-2035.
[3] Su V, Lau AF. Connexins: mechanisms regulating protein levels and intercellular communication. FEBS Lett 2014; 588(8): 1212-20.
[4] Harris AL, Locke D. Permeability of connexin channels. In: Haris AL, Locke D, eds. Connexins: A Guide. Humana Press, New York 2008; pp. 165-206.
[5] Sáez JC, Leybaert L. Hunting for connexin hemichannels. FEBS Lett 2014; 588(8): 1205-11. Review.
[6] Retamal MA, Reyes EP, García IE, *et al.* Diseases associated with leaky hemichannels. Front Cell Neurosci. 2015; 9: 267. Review.
[7] Rackauskas M1, Neverauskas V, Skeberdis VA. Diversity and properties of connexin gap junction channels. Medicina (Kaunas) 2010; 46(1): 1-12.
[8] Kelly JJ, Simek J, Laird DW. Mechanisms linking connexin mutations to human diseases. Cell Tissue Res 2015; 360(3): 701-21.
[9] Xu J, Nicholson BJ. The role of connexins in ear and skin physiology - functional insights from disease-associated mutations. Biochim Biophys Acta 2013; 1828(1): 167-78.
[10] Denoyelle F, Weil D, Maw MA, *et al.* Prelingual deafness: high prevalence of a 30delG mutation in the connexin 26 gene. Hum Mol Genet 1997; 6(12): 2173-7.
[11] Kelsell DP, Dunlop J, Stevens H-P, *et al.* Connexin 26 mutations in hereditary non-syndromic sensorineural deafness. Nature 1997; 1, 387(6628): 80-3.
[12] Zelante L, Gasparini P, Estivill X, *et al.* Connexin26 mutations associated with the most common form of non-syndromic neurosensory autosomal recessive deafness (DFNB1) in Mediterraneans. Hum Mol Genet 1997; 6(9): 1605-9.
[13] Melchionda S, Bicego M, Marciano E, *et al.* Functional characterization of a novel Cx26 (T55N) mutation associated to non-syndromic hearing loss. Biochem Biophys Res Commun 2005; 337(3): 799-805.
[14] Beyer EC, Lipkind GM, Kyle JW, *et al.* Structural organization of intercellular channels II. Amino terminal domain of the connexins: sequence, functional roles, and structure. Biochim Biophys Acta 2012; 1818(8): 1823-30.
[15] Terrinoni A, Codispoti A, Serra V, *et al.* Connexin 26 (GJB2) mutations as a cause of the KID syndrome with hearing loss. Biochem Biophys Res Commun 2010; 395(1): 25-30.
[16] García IE, Maripillán J, Jara O, *et al.* Keratitis-ichthyosis-deafness syndrome-associated cx26 mutants produce nonfunctional gap junctions but hyperactive hemichannels when co-expressed with wild type cx43. J Invest Dermatol 2015; 135(5): 1338-47.
[17] Liu W, Boström M, Kinnefors A, et al. Unique expression of connexins in the human cochlea. Hear Res 2009; 250(1-2): 55-62.
[18] Adachi N, Yoshida T, Nin F, *et al.* The mechanism underlying maintenance of the endocochlear potential by the K+ transport system in fibrocytes of the inner ear. J Physiol 2013; 591: 4459-72.

[19] Jagger DJ, Forge A. Connexins and gap junctions in the inner ear--it's not just about K^+ recycling. Cell Tissue Res 2015; 360(3): 633-44

[20] Avshalumova L, Fabrikant J, Koriakos A. Overview of skin diseases linked to connexin gene mutations. Int J Dermatol. 2014; 53(2): 192-205.

[21] Langlois S, Maher AC, Manias JL, *et al*. Connexin levels regulate keratinocyte differentiation in the epidermis. J Biol Chem 2007; 282(41): 30171-80.

[22] Scott CA, Tattersall D, O'Toole EA, *et al*. Connexins in epidermal homeostasis and skin disease. Biochim Biophys Acta 2012; 1818(8): 1952-61. Review.

[23] Wang X, Ramirez A, Budunova I. Overexpression of connexin26 in the basal keratinocytes reduces sensitivity to tumor promoter TPA. Exp Dermatol 2010; 19(7): 633-40.

[24] Shuja Z, Li L, Gupta S, Meşe G, *et al*. Connexin26 mutations causing palmoplantar keratoderma and deafness interact with connexin43, modifying gap junction and hemichannel properties. J Invest Dermatol 2015 Oct 6. doi: 10.1038/jid.2015.389. [Epub ahead of print]

[25] Seirafi H1, Khezri S, Morowati S, *et al*. A new variant of Vohwinkel syndrome: a case report. Dermatol Online J 2011; 17(3): 3.

[26] Iossa S, Chinetti V, Auletta G, *et al*. New evidence for the correlation of the p.G130V mutation in the GJB2 gene and syndromic hearing loss with palmoplantar keratoderma. Am J Med Genet A 2009; 15, 149A(4): 685-8.

[27] Pang X, Chai Y, Sun L, *et al*. Characterization of spectrum, *de novo* rate and genotype-phenotype correlation of dominant GJB2 mutations in Chinese hans. PLoS One 2014; 9(6): e100483.

[28] Bondeson ML, Nyström AM, Gunnarsson U, *et al*. Connexin 26 (GJB2) mutations in two Swedish patients with atypical Vohwinkel (mutilating keratoderma plus deafness) and KID syndrome both extensively treated with acitretin. Acta Derm Venereol 2006; 86(6): 503-8.

[29] Jonard L, Feldmann D, Parsy C, *et al*. A familial case of Keratitis-Ichthyosis-Deafness (KID) syndrome with the GJB2 mutation G45E. Eur J Med Genet 2008; 51(1): 35-43.

[30] Sbidian E, Feldmann D, Bengoa J, Fraitag S, *et al*. Germline mosaicism in keratitis-ichthyosis-deafness syndrome: pre-natal diagnosis in a familial lethal form. Clin Genet 2010; 77(6): 587-92.

[31] Todt I, Mazereeuw-Hautier J, Binder B, *et al*. Dandy-Walker malformation in patients with KID syndrome associated with a heterozygote mutation (p.Asp50Asn) in the GJB2 gene encoding connexin 26. Clin Genet 2009; 76(4): 404-8.

[32] Lazic T, Li Q, Frank M, *et al*. Extending the phenotypic spectrum of keratitis-ichthyosis-deafness syndrome: report of a patient with GJB2 (G12R) Connexin 26 mutation and unusual clinical findings. Pediatr Dermatol 2012; 29(3): 349-57.

[33] Mayama H, Fujimura T, Asano M, *et al*. Squamous cell carcinoma arising from Keratitis-ichthyosis-deafness syndrome. Acta Derm Venereol 2013; 93(5): 583-4.

[34] Coggshall K, Farsani T, Ruben B, *et al*. Keratitis, ichthyosis, and deafness syndrome: a review of infectious and neoplastic complications. J Am Acad Dermatol 2013; 69(1): 127-34. Review.

[35] van Geel M, van Steensel MA, Küster W, *et al*. HID and KID syndromes are associated with the same connexin 26 mutation. Br J Dermatol 2002; 146(6): 938-42.

[36] Todt I1, Hennies HC, Küster W, *et al*. Neurotological and neuroanatomical changes in the connexin-26-related HID/KID syndrome. Audiol Neurootol 2006; 11(4): 242-8.

[37] Richard G, Brown N, Ishida-Yamamoto A, *et al*. Expanding the phenotypic spectrum of Cx26 disorders: Bart-Pumphrey syndrome is caused by a novel missense mutation in GJB2. J Invest Dermatol 2004; 123(5): 856-63.

[38] Braun-Falco M. Hereditary palmoplantar keratodermas. J Dtsch Dermatol Ges 2009; 7(11): 971-84.

[39] Bart RS, Pumphrey RE. Knuckle pads, leukonychia and deafness. A dominantly inherited syndrome. N Engl J Med 1967; 276(4): 202-7.

[40] Fiori MC, Krishnan S, Cortes DM, *et al*. Functional hemichannels formed by human connexin 26 expressed in bacteria. Biosci Rep 2015; 35(2).

[41] Qu Y, Tang W, Zhou B, *et al*. Early developmental expression of connexin26 in the cochlea contributes to its dominate functional role in the cochlear gap junctions. Biochem Biophys Res Commun 2012; 417(1): 245-50.

[42] Mani RS, Ganapathy A, Jalvi R, *et al*. Functional consequences of novel connexin 26 mutations associated with hereditary hearing loss. Eur J Hum Genet 2009; 17(4): 502-9.

[43] Zhang J, Scherer SS, Yum SW. Dominant Cx26 mutants associated with hearing loss have dominant-negative effects on wild type Cx26. Mol Cell Neurosci 2011; 47(2): 71-8.

[44] Mese G, Sellitto C, Li L, *et al.* The Cx26-G45E mutation displays increased hemichannel activity in a mouse model of the lethal form of keratitis-ichthyosis-deafness syndrome. Mol Biol Cell 2011; 22(24): 4776-86.

[45] Mhaske PV, Levit NA, Li L, *et al.* The human Cx26-D50A and Cx26-A88V mutations causing keratitis-ichthyosis-deafness syndrome display increased hemichannel activity. Am J Physiol Cell Physiol 2013; 304(12): C1150-8.

[46] Meigh L, Hussain N, Mulkey DK, *et al.* Connexin26 hemichannels with a mutation that causes KID syndrome in humans lack sensitivity to CO2. Elife 2014; 3: e04249.

[47] Levit NA, Sellitto C, Wang HZ, *et al.* Aberrant connexin26 hemichannels underlying keratitis-ichthyosis-deafness syndrome are potently inhibited by mefloquine. J Invest Dermatol 2015; 135(4): 1033-42.

[48] Yoshikawa S, Kawano A, Hayashi, C, *et al.* The clinical features of patients with the homozygous 235delC and the compound-heterozygous Y136X/G45E of the GJB2 mutations (Connexin 26) in cochlear implant recipients. Auris Nasys Larynx 2011; 38(4): 444-9.

[49] Snoeckx RL, Hassan DM, Kamal NM, *et al.* Mutation analysis of the GJB2 (connexin 26) gene in Egypt. Hum Mutat 2005; 26(1): 60-1.

[50] Sanchez HA, Bienkowski R, Slavi N, *et al.* Altered inhibition of Cx26 hemichannels by pH and Zn2+ in the A40V mutation associated with keratitis-ichthyosis-deafness syndrome. J Biol Chem 2014; 289(31): 21519-32.

[51] Richard G, Smith LE, Bailey RA, *et al.* Mutations in the human connexin gene GJB3 cause erythrokeratodermia variabilis. Nat Genet 1998; 20(4): 366-9.

[52] Zhang XJ, Chen JJ, Yang S, *et al.* A mutation in the connexin 30 gene in Chinese Han patients with hidrotic ectodermal dysplasia. J Dermatol Sci 2003; 32(1): 11-7.

[53] van Steensel MA, Oranje AP, van der Schroeff JG, *et al.* The missense mutation G12D in Connexin30.3 can cause both erythrokeratodermia variabilis of Mendes da Costa and progressive symmetric erythrokeratodermia of Gottron. Am J Med Genet A 2009; 149A(4): 657-61.

[54] Tattersall D, Scott CA, Gray C, *et al.* EKV mutant connexin 31 associated cell death is mediated by ER stress. Hum Mol Genet 2009; 18(24): 4734-45.

[55] Maeda S, Nakagawa S, Suga M, *et al.* Structure of the connexin 26 gap junction channel at 3.5 A resolution. Nature. 2009; 458(7238): 597-602.

[56] Nakagawa S, Maeda S, Tsukihara T. Structural and functional studies of gap junction channels. Curr Opin Struct Biol 2010; 20(4): 423-30.

[57] Lynn BD, Tress O, May D, *et al.* Ablation of connexin30 in transgenic mice alters expression patterns of connexin26 and connexin32 in glial cells and leptomeninges. Eur J Neurosci 2011; 34(11): 1783-93.

[58] Lee JR, Derosa AM, White TW. Connexin mutations causing skin disease and deafness increase hemichannel activity and cell death when expressed in Xenopus oocytes. J Invest Dermatol 2009; 129(4): 870-8.

[59] Sánchez HA, Mese G, Srinivas M, *et al.* Differentially altered Ca2+ regulation and Ca2+ permeability in Cx26 hemichannels formed by the A40V and G45E mutations that cause keratitis ichthyosis deafness syndrome. J Gen Physiol 2010; 136(1): 47-62.

[60] Levit NA, Mese G, Basaly MG, *et al.* Pathological hemichannels associated with human Cx26 mutations causing Keratitis-Ichthyosis-Deafness syndrome. Biochim Biophys Acta 2012; 1818(8): 2014-9.

[61] Lopez W, Gonzalez J, Liu Y, Harris AL *et al.* Insights on the mechanisms of Ca(2+) regulation of connexin26 hemichannels revealed by human pathogenic mutations (D50N/Y). J Gen Physiol 2013; 142(1): 23-35.

[62] Akiyama M, Sakai K, Arita K, *et al.* A novel GJB2 mutation p.Asn54His in a patient with palmoplantar keratoderma, sensorineural hearing loss and knuckle pads. J Invest Dermatol 2007; 127(6): 1540-3.

[63] Yum SW, Zhang J, Scherer SS. Dominant connexin26 mutants associated with human hearing loss have trans-dominant effects on connexin30. Neurobiol Dis 2010; 38(2): 226-36.

[64] de Zwart-Storm EA, van Geel M, Veysey E, *et al.* A novel missense mutation in GJB2, p.Tyr65His, causes severe Vohwinkel syndrome. Br J Dermatol 2011; 164(1): 197-9.

[65] de Zwart-Storm EA, Hamm H, Stoevesandt J, *et al.* A novel missense mutation in GJB2 disturbs gap junction protein transport and causes focal palmoplantar keratoderma with deafness. J Med Genet 2008; 45(3): 161-6.

CHAPTER 3

Tyrosine Kinase Inhibitors; Therapies for Thyroid Cancer

Silvia Martina Ferrari[1], Poupak Fallahi[1], Enke Baldini[2], Salvatore Ulisse[2], Gabriele Materazzi[3], Paolo Miccoli[3] and Alessandro Antonelli[1,*]

[1]Department of Clinical and Experimental Medicine, University of Pisa, Via Savi, 10, 56126, Pisa, Italy; [2]Department of Experimental Medicine, "Sapienza" University of Rome, Viale dell'Università 30, 00185 Rome, Italy and [3]Department of Surgical, Medical, Molecular Pathology and Critical Area, University of Pisa, Via Savi, 10, 56126, Pisa, Italy

Abstract: In spite of the good prognosis of thyroid carcinoma (TC), a metastatic disease is developed by approximately 5% of patients, not responsive to radioactive iodine (RAI), and with a more aggressive behaviour. The absence of specific and effective drugs for aggressive TC leads to the need of new efforts towards new drugs development.

Several genetic alterations in different molecular pathways in TC have been shown in the last decades, associated with TC development and progression. Rearranged during transfection (RET)/papillary thyroid carcinoma gene rearrangements, RET mutations, BRAF mutations, RAS mutations, and vascular endothelial growth factor receptor 2 angiogenesis pathways are some of the studied pathways, determinant in TC development. Tyrosine kinase inhibitors (TKIs) are small organic compounds inhibiting tyrosine kinases auto-phosphorylation and activation, most of them are multikinase inhibitors. TKIs act on the above-mentioned molecular pathways involved in growth, angiogenesis, local and distant spread of TC. TKIs are emerging as new therapies of aggressive TC, including differentiated thyroid cancer (DTC), medullary thyroid cancer (MTC) and anaplastic thyroid cancer (ATC), and they have been shown capable of inducing clinical responses and stabilization of disease. Vandetanib and cabozantinib have been approved for MTC treatment; sorafenib and lenvatinib have been approved for DTC refractory to RAI. These drugs prolong median progression-free survival, but until now no significant increase has been observed on overall survival; side effects are common. New efforts are made to find new more effective and safe compounds, and to personalize the therapy in each TC patient.

Keywords: Anaplastic thyroid cancer, BRAF mutations, cabozantinib, CLM3, CLM94, differentiated thyroid cancer, follicular thyroid cancer, H-RAS mutations, K-RAS mutations, lenvantinib, medullary thyroid cancer, N-RAS mutations, papillary thyroid cancer, pyrazolpyrimidine, Rearranged during

***Corresponding author Alessandro Antonelli:** Department of Clinical and Experimental Medicine, University of Pisa, Via Savi, 10, I-56126, Pisa, Italy; E-mail: alessandro.antonelli@med.unipi.it

Atta-ur-Rahman (Ed.)

transfection (RET)/papillary thyroid carcinoma gene rearrangements, RET mutations, sorafenib, telomerase reverse transcriptase (TERT) mutations, vandetanib, vascular endothelial growth factor receptor 2.

INTRODUCTION

The incidence of papillary thyroid cancer (PTC) is increasing in the last decades [1-4], and it is the predominant endocrine carcinoma (7th most common cause of new cancer in the US for women).

The use of neck ultrasonography and fine needle aspiration (FNA) of thyroid nodules is related to the increasing diagnosis of small thyroid cancer (TC). However other reasons of this increasing incidence are not clear. Ionizing radiations play an important role as risk factor of TC, in particular if the exposure is in the early life (2-4% of subjects exposed to radiation during childhood show TC) [5]. The exposure to fall-out of nuclear explosions or accidents constitutes a risk factor for TC [6]; it is well known that the incidence of TC in childhood rapidly increased, after the Chernobyl nuclear accident in the regions exposed to the radiation fallout of Belarus and Ukraine [7]. Low doses of radiation exposure might also be related to TC [8]. A higher frequency of follicular thyroid cancer (FTC) has been observed in iodine deficient areas [9]. Moreover, autoimmune thyroiditis has been shown to be a risk factor for PTCs [10, 11]. Recently, new risk factors are emerging [12].

According to histopathological criteria: 1- differentiated thyroid carcinomas (DTC) that arise from follicular cells account >90% of TC, and are subdivided into PTC (about 90%) or FTC (about 10%); 2- medullary thyroid cancer (MTC) accounts < 5% of TC [1] sporadic (75%); or hereditary forms (25%) that include familial MTC (FMTC), MEN 2A, MEN 2B [13, 14]); 3- anaplastic thyroid cancer (ATC) accounts for less than 2%.

Completion total thyroidectomy is the treatment of choice for PTCs, FTCs and MTC [15]. Radioiodine should be routinely recommended in high-risk (HR) patients and considered in intermediate risk (IR) DTC patients [16, 17].

Patients with PTC and FTC (after surgery, and eventually radioiodine) are followed by neck ultrasonography and the determination of basal or thyroid-stimulating hormone (TSH)-stimulated thyroglobulin (Tg) [18-20].

More than 90% of patients with DTC has a normal life expectancy [21]. The prevalence of TC deaths per year (in relation to the number of new cases) is decreased from 15% to 5% in the last decades [13], in contrast with the increasing incidence.

About 5% of DTC patients have distant metastasis at the diagnosis; in the follow-up, from 10% to 15% of patients have recurrent disease, and survival diminishes from 70% to 50%, in 10 years [22, 23].

During tumor progression, the iodide uptake ability may be lost by DTC cancer cells, becoming not responsive to radioiodine, with a significant worsening of the prognosis [21-23].

Radiotherapy and chemotherapy (doxorubicin) are of limited efficacy in dedifferentiated thyroid cancer (DePTC) treatment [24].

MOLECULAR PATHWAYS IN DTC PROGRESSION

Several somatic mutations in different molecular pathways in TC have been shown during the last decades; they have been associated with TC development and progression.

RET (REarranged during Transfection) is a proto-oncogene encoding a transmembrane protein harboring a tyrosine kinase (TK). Mutations and rearrangements that activate RET have been identified in various human cancers [25]. In about 40% of adult sporadic PTC, RET/PTC rearrangements are present [26]; RET/PTC1, and RET/PTC3 are the most common, and they are reported in microcarcinomas and, also, in benign thyroid lesions: for these reasons it has been hypothesized that RET/PTC are determinant for tumor initiation, but not progression [27-30].

The BRAF kinase is a member of the RAF family proteins; a BRAF-V600E mutation is present in about 45% of PTC, and it has been associated with tumor aggressiveness [26].

Mutations of the promoter of the telomerase reverse transcriptase (TERT) gene are found in 11% PTCs, 11% of FTCs, 37% of poorly differentiated thyroid cancers (PDTC), and 42% of ATCs, and are particularly prevalent in the aggressive TCs (overall in the presence of BRAF mutations) [31, 32].

Intracellular G-proteins are encoded by K-RAS, N-RAS and H-RAS (belong to the RAS gene family), taking part in the activation of intracellular signaling pathways. RAS mutations are present in ~10% of PTCs, and in ~40-50% of FTCs. RAS mutations are linked to a more aggressive TC behaviour [33].

PAX8/PPARγ rearrangements are present in ~30-40% of FTC, ~2-10% of follicular adenomas, and less frequently in the follicular variant of PTC [26].

An increase in angiogenesis and a more aggressive TC behavior correlates each other [34], and the expression of angiogenesis inhibitors or stimulators (Vascular Endothelial Growth Factor (VEGF)/VEGF receptor (VEGFR), Epidermal Growth Factor (EGF)/EGFR, platelet-derived growth-factor (PDGF)/PDGFR, fibroblast growth factor (FGF)/FGFR, and hepatocyte growth factor (HGF)/c-Met) in TC is associated with clinical features of the disease [35]. VEGF is overexpressed (such as VEGFR-2) in DTC [36] and it takes part in neoplastic progression and aggressiveness. The dispensing of drugs targeting VEGF pathway is actually a therapeutic option for TC patients [37].

Also EGFR is highly expressed in aggressive PTC in presence of lymph node metastasis, and takes part in the progression of TC [38, 39].

MOLECULAR PATHWAYS IN MTC

MTC is a slow growing cancer, however the 10-year overall survival (OS) rate is 40–50% in the presence of metastatic disease. Surgery is the only curative modality [40]; metastatic MTC is not responsive to external radiotherapy, or chemotherapy.

The major part of MTC cases are sporadic; hereditary forms (25%) include familial MTC (FMTC), MEN 2A, MEN 2B [14].

Almost all hereditary cases of MTC and 30–50% of the sporadic ones show activating RET mutations [41], so RET is a determinant therapeutic target for MTC [42].

Recent findings have shown that H-RAS mutations are present in 13-56% of RET-negative sporadic MTC, and the activation of the mammalian target of rapamycin (mTOR) intracellular signaling pathway in hereditary MTC [43-45].

TYROSINE KINASE INHIBITORS (TKIS)

TKIs are usually small organic compounds that inhibit TK auto-phosphorylation and activation, carrying out a competition with the ATP-binding site of the catalytic domain of the TKs [46]. A single TKI may target one or multiple TKs [47]; however, most of them are multikinase inhibitors. TKIs act on the above-mentioned molecular pathways involved in growth, angiogenesis, local and distant spread [48].

TKIs are emerging as new therapies of aggressive TC, including DTC, MTC and ATC [49, 50]. Some TKIs have been approved for the treatment of different types of cancer, as imatinib (for chronic myelogenous leukemia [51] and gastrointestinal stromal tumors [52]), sorafenib (for advanced renal cell carcinoma (RCC) [53], advanced hepatocellular carcinoma (HCC) [54], and radioactive iodine (RAI) resistant advanced TC [55]), vandetanib and cabozantinib for MTC [56-59], and others [60].

Sorafenib

Sorafenib (Nexavar®) is (a bi-aryl urea) multi-targeted TKI, with inhibitor activity against VEGFR-2 and 3, c-Kit, PDGFR, RET/PTC, and Raf kinases (more avidly C-Raf than B-Raf), and the Raf/Mek/Erk pathway (MAPK pathway); it has been also shown to induce apoptosis through down-regulation of Mcl-1 [55, 61].

Sorafenib is approved for the treatment of primary kidney cancer (advanced RCC) and advanced primary liver cancer (HCC).

According to different phase II trials [62-67], it has been conducted a multicenter (DECISION trial), double-blind randomized phase III study, in locally advanced/metastatic RAI-refractory DTC (a target lesion without iodine uptake on a post RAI scan performed during a low iodine diet and adequate TSH elevation or recombinant human TSH stimulation) *versus* placebo [68]. Patients have been randomized 1:1 to receive placebo or sorafenib. The initial group comprises a population of 417 patients (207 treated with sorafenib and 210 with placebo), while the final group was constituted by 416 patients (207 treated with sorafenib and 209 with placebo). The inclusion criteria were: age >18, life expectancy not fewer than 12 weeks, locally advanced/metastatic DTC (PTC, FTC, Hurtle cell or PDTC) with at least one measurable lesion (by computed tomography or magnetic resonance imaging) and a progression of disease by 14 months. Other inclusion criteria were: a performance status < 2 according to Eastern Cooperative

Oncology Group, adequate TSH suppression (<0.5 mU/L), absence of renal and liver failure and adequate bone marrow function. Patients were excluded, if previously treated with TKI, monoclonal antibodies against VEGFRs or other targeted agents, cytotoxic chemotherapy or thalidomide. Efficacy and safety of sorafenib have been assessed every 56 days (two cycles) and 28 days (1 cycle) for the first 8 months and every 56 days after, respectively. At the end of the study, median progression-free survival (PFS) was significantly higher in patients receiving sorafenib (10.8 months) than in the ones receiving placebo (5.8 months), improving in the pre-specified clinical and genetic biomarker subgroups, independently from mutation status [68].

A phase III trial, conducted by Bayer, involving 417 patients, is still ongoing [69].

Vandetanib

Vandetanib (Caprelsa, ZD6474, AstraZeneca) is a multi-targeted TKI, with inhibitor activity against RET, VEGFR-2, VEGFR-3 and EGFR and to a lesser extent VEGFR-1 [70].

Vandetanib is able to block the enzymatic activity of RET-derived oncoproteins, the proliferation of two human PTC cell lines carrying spontaneous RET/PTC1 rearrangements, and the growth of tumors *in vivo* upon the injection of RET/PTC3-transformed NIH3T3 fibroblasts into nude mice [71, 72]. It was also shown that vandetanib inhibited the growth of a transplantable MTC (from a sporadic human MTC carrying a RET C634R mutation) in nude mice [73].

However, *in vivo*, the most important antitumoral effect of vandetanib is an inhibition of angiogenesis, that is exerted interfering with EGFR-induced production of angiogenic growth factors [74, 75].

Several clinical trials have evaluated the efficacy of vandetanib in patients with TC.

A randomized, double-blind, placebo-controlled (ZETA trial) multicentre phase III trial was conducted in 331 MTC patients [advanced (5%) or metastatic (95%) [76]; 10% with hereditary MTC, 90% sporadic MTC; 56% of patients were RET mutation positive, 2% RET mutation negative and 41% unknown]. All MTC patients were offered the option of receiving vandetanib (300 mg daily administered until disease progression) in an open-label phase. Vandetanib prolonged PFS significantly than placebo (30.5 *versus* 19.3 months; $p<0.001$), while OS was not significantly changed; calcitonin (69% *versus* 3%) and CEA (52% *versus* 2%) response rate were also higher ($p<0.001$) in patients receiving vandetanib. The response to therapy was good also in patients with sporadic MTC (RET positive), in patients with M918T negative tumors and in RET unknown status [76].

On the basis of these results, vandetanib was approved, since April (FDA) and November (EU) 2011, for the treatment of unresectable, locally advanced or metastatic MTC patients, in Europe and USA [77].

A recent retrospective study has shown the efficacy of vandetanib in the treatment of 11 progressive metastatic MTC patients with 36% partial response (PR) [67], too.

Furthermore, another study showed vandetanib efficacy in advanced MTC patients, but it was observed the appearance of at least one adverse event (AE) in all patients, and a careful monitoring was recommended [78].

A first phase IV randomized trial is ongoing to evaluate differences in the response in MTC patients treated with vandetanib 300 mg *versus* 150 mg/day [79].

A new randomized, double-blind, placebo-controlled, multi-centre phase III study is actually ongoing in DTC patients [80].

Cabozantinib

Cabozantinib [(Cometriq, XL184, Exelixis), known variously as N-(4-{[6,7-bis(methyloxy)quinolin-4-yl]oxy}phenyl)-N'-(4-fluorophenyl)cyclopropane-l,l-dicarboxamide (WO2014165786)] is a multitarget TKI that inhibits MET, VEGFR-2, RET, KIT (mast/stem cell growth factor), AXL, FLT3 (FMS-like tyrosine kinase 3) and Tie-2 (tunica interna endothelial cell kinase 2), that are involved in angiogenesis and cell proliferation [81, 82]. All these TKs are important in the development and progression of TC [58]; inhibiting TKs, cabozantinib avoids receptors phosphorylation and thus blocks cell proliferation, angiogenesis, growth and invasiveness of tumors [83, 84].

A recent *in vitro* study has shown that cabozantinib inhibited cell proliferation in PTC cells harboring RET/PTC1 rearrangement, enhancing the expression of iodide-handling genes and inhibiting the expression of glucose transporter genes [61].

An international, double-blind, randomized trial (EXAM), was conducted in 330 patients [with an age > 65 years or ≤ 65 years) and prior TKIs therapy] who showed radiographic progression of metastatic MTC in the previous 14 months [85], comparing 140 mg/day of cabozantinib *versus* placebo. In about 50% of patients, RET mutations were present, mainly M918T (74%), while mutation status was unknown in 39% of patients. PFS (primary endpoint) was 11.2 months in the patients receiving cabozantinib *versus* 4.0 months in the ones receiving placebo ($p<0.001$) with an improvement of PFS independent of age, bone metastases, RET status and prior treatment. Patients alive and progression-free at 1 year were 47.3% for cabozantinib, and 7.2% for placebo, according to Kaplan-Meier estimator. In the cabozantinib group of patients, response rate (partial) was 28% (duration of response of 14.6 months), and it was similar in RET + and RET- patients. A reduction of the lesion size was observed in 94% of patients in the cabozantinib arm, *versus* 27% in the placebo arm. Moreover, a significant decrease of calcitonin and CEA levels were observed in the cabozantinib treated patients, that was correlated to the reduction of the size of target lesions [85].

Cabozantinib was approved in November 2012 by FDA for metastatic MTC and in December 2013 by EMA for progressive, locally advanced or metastatic MTC [86, 87].

Two oral formulations exist: capsules (CN103830203) and dispersible tablets. These latter, compared with common tablets, have high solubility, bioavailability, rapid in *in vivo* distribution, are stable in quality, good in taste (CN103751140) [88, 89]. The advisable cabozantinib dose is 140 mg per day (1 capsule of 80 mg and 3 of 20 mg), and patients should not eat 2 hours before and at least 1 hour after the therapy [59].

Lenvatinib

Lenvatinib is an oral, multitargeted TKI of the VEGFRs 1, 2, and 3, FGFRs 1 through 4, PDGFR α, RET, and KIT [90, 91]. According to the data shown by a phase II study conducted in patients with aggressive TC [92], a phase III study on lenvatinib (E7080) has been carryed out in Differentiated Cancer of the Thyroid (SELECT) in TC patients resistant to iodine-131 [93]. In this phase III study, in progressive TC resistant to iodine-131, 261 patients received lenvatinib (one daily dose of 24 mg in cycles of 28 days), and 131 received placebo. If disease progressed, patients receiving placebo could be also administered with open-label lenvatinib. The primary end point was PFS, while secondary ones were response rate, OS, and safety. Lenvatinib prolonged significantly median PFS (18.3

months) than placebo (3.6 months) (p<0.001). The response rate was 64.8% in patients receiving lenvatinib (4 complete responses, 165 PR), *versus* 1.5% in the ones receiving placebo (p<0.001), while median OS did not differ significantly among them. More than 40% of patients administered with lenvatinib showed AEs: hypertension (in 67.8% of patients), diarrhea (59.4%), fatigue or asthenia (59.0%), decreased appetite (50.2%), decreased weight (46.4%), and nausea (41.0%). Discontinuations happened in 37 patients receiving lenvatinib (14.2%) and 3 receiving placebo (2.3%) owing to AEs. Among the 20 deaths during the treatment period, drug-related deaths were in 6 patients treated with lenvatinib [93].

Pyrazolpyrimidine

One of the most explored chemical template is the pyrazolo[3,4-d]pyrimidine (PP) heterocyclic core, a useful scaffold for the obtainment of effective TKIs. Actually, derivatives belonging to this structural class show a large spectrum of activity. Different PP compounds act as: a) Abl inhibitors and antiproliferative agents against human leukemia cell lines; b) Src kinase inhibitors in neuroblastoma, medulloblastoma and osteosarcoma; c) Phospholipase D inhibitors in different neoplasias; d) Urokinase plasminogen activator inhibitors, in breast cancer.

Recently CLM3, [(R)-1-phenethyl-N-(1-phenylethyl)-1H-pyrazolo[3,4-d]pyrimidin-4-amine], has been shown to inhibit RET-TK, BRAF, VEGFR-2, EGFR, and to exert antiangiogenic activity. In human TC cell lines CLM3 showed antiproliferative and proapoptotic effects and also an antiangiogenic effect [94-96].

It has been also shown that CLM3 and CLM29 (another pyrazolo[3,4-d]pyrimidine, inhibiting RET, EGFR, VEGFR, and with anti-angiogenic activity) [94] inhibited the migration of DePTC cells. Moreover, CLM3 [97] and CLM29 [94], have been demonstrated to exert antineoplastic activity in primary ATC cells [98]. The inhibitory effect of CLM3 and CLM29 did not depend on the presence/absence of BRAF-V600E mutation both in DePTc and in ATC. A DePTC cell line (AL), with BRAF-V600E mutation, was produced, which was able to grow in nu/nu mice when inoculated sc. CLM3 and CLM29 increased TSP-1 expression in the AL cell line. The antineoplastic activity of CLM3 and CLM29 could depend on the antiproliferative effect, the apoptosis increase in tumoral cells and the inhibition of migration and neoplastic neovascularization. In fact, a significant reduction of microvessels was observed, *in vivo*, in the CLM3-treated tumors [99, 100].

It has been recently [101] reported the *in vitro* antineoplastic and antiangiogenic effects of CLM3 and CLM94 in primary cell cultures of MTC (pMTC) and in the MTC cell lines, TT (with the RET C634W activating mutation) and MZ-CRC-1 (with the MEN2B RET mutation Met891Thr). The inhibitory effect of CLM3 and CLM94 resulted not dependent from the presence/absence of RET mutation in pMTC. VEGF-A gene expression decreased significantly in TT cells treated with CLM3 and CLM94.

More recently, CLM29 was tested both in pMTC cells established from surgical biopsies, and in TT cells with the C634W RET mutation [102]. CLM29 inhibited significantly the proliferation (similarly in pMTC cells with/without RET mutation), and increased the apoptosis of pMTC and TT cells. Upon the injection of TT cells sc in CD nu/nu mice, tumor masses became detectable after 20-30 days; CLM29 (50 mg/kg/die) significantly reduced tumor growth and weight, and microvessel density. These data demonstrated the antineoplastic activity of CLM29 in MTC *in vitro*, and *in vivo*, enabling an eventual clinical evaluation [102].

CONCLUSION

In spite of the good prognosis of thyroid carcinoma (TC), a metastatic disease is developed by approximately 5% of patients, not responsive to radioactive iodine (RAI), and with a more aggressive behaviour. The absence of specific and effective drugs for aggressive TC leads to the need of new efforts towards new drugs development.

Several genetic alterations in different molecular pathways in TC have been shown in the last decades, that have been associated with TC development and progression. RET/PTC gene rearrangements, RET mutations, BRAF mutations, RAS mutations, and VEGFR-2 angiogenesis pathways are some of the studied pathways determinant in TC development. TKIs are usually small organic compounds that inhibit TK auto-phosphorylation and activation, most of them are multikinase inhibitors. TKIs act on the above-mentioned molecular pathways involved in growth, angiogenesis, local and distant spread of TC. TKIs are emerging as new therapies of aggressive TC, including DTC, MTC and ATC, and they have been shown capable of inducing clinical responses and stabilization of disease. Vandetanib and cabozantinib have been approved for the treatment of MTC. Furthermore, sorafenib and lenvatinib have been approved for DTC refractory to RAI. These drugs prolong median PFS, but until now no significant increase has been observed on OS; side effects are common. New efforts are made to find new more effective and safe compounds, and to personalize the therapy in each TC patient.

ACKNOWLEDGEMENTS

Declared none.

CONFLICT OF INTEREST

The author confirms that he has no conflict of interest to declare for this publication.

DISCLOSURE

This is an updated chapter of the earlier review article [by A. Antonelli "New targeted therapies for thyroid cancer". *Curr Genomics.* 2011 Dec;12(8):626-31], with additional text and references.

REFERENCES

[1] Cancer statistics highlight progress, challenges. Cancer Discov 2014; 4: OF3.
[2] Colonna M, Uhry Z, Guizard A, *et al.* Recent trends in incidence, geographical distribution, and survival of papillary thyroid cancer in France. Cancer Epidemiol 2015; (in press).
[3] Finlayson A, Barnes I, Sayeed S, McIver B, Beral V, Ali R. Incidence of thyroid cancer in England by ethnic group, 2001-2007. Br J Cancer 2014; 110(5): 1322-7.
[4] Enewold LR, Zhou J, Devesa SS, *et al.* Thyroid cancer incidence among active duty U.S. military personnel, 1990-2004. Cancer Epidemiol Biomarkers Prev 2011; 20(11): 2369-76.
[5] Schlumberger M, Chevillard S, Ory K, Dupuy C, Le Guen B, de Vathaire F. Thyroid cancer following exposure to ionizing radiation. Cancer Radiother 2011; 15(5): 394-9.
[6] Braverman ER, Blum K, Loeffke B, *et al.* Managing terrorism or accidental nuclear errors, preparing for iodine-131 emergencies: a comprehensive review. Int J Environ Res Public Health 2014; 11(4): 4158-200.
[7] Antonelli A, Miccoli P, Derzhitski VE, Panasiuk G, Solovieva N, Baschieri L. Epidemiologic and clinical evaluation of thyroid cancer in children from the Gomel region (Belarus). World J Surg 1996; 20(7): 867-71.
[8] Suzuki K, Yamashita S. Low-dose radiation exposure and carcinogenesis. Jpn J Clin Oncol 2012; 42(7): 563-8.
[9] Zimmermann MB, Boelaert K. Iodine deficiency and thyroid disorders. Lancet Diabetes Endocrinol 2015; 3(4): 286-95.
[10] Jankovic B, Le KT, Hershman JM. Clinical Review: Hashimoto's thyroiditis and papillary thyroid carcinoma: is there a correlation? J Clin Endocrinol Metab 2013; 98(2): 474-82.
[11] Antonelli A, Ferri C, Fallahi P, *et al.* Clinical and subclinical autoimmune thyroid disorders in systemic sclerosis. Eur J Endocrinol 2007; 156(4): 431-7.
[12] Antonelli A, Ferri C, Fallahi P, *et al.* Thyroid cancer in HCV-Related chronic hepatitis patients: A case-control study. Thyroid 2007; 17(5): 447-51.
[13] Ward EM, Thun MJ, Hannan LM, Jemal A. Interpreting cancer trends. Ann N Y Acad Sci 2006; 1076: 29-53.
[14] Pusztaszeri MP, Bongiovanni M, Faquin WC. Update on the cytologic and molecular features of medullary thyroid carcinoma. Adv Anat Pathol 2014; 21(1): 26-35.

[15] Miccoli P, Antonelli A, Spinelli C, Ferdeghini M, Fallahi P, Baschieri L. Completion total thyroidectomy in children with thyroid cancer secondary to the Chernobyl accident. Arch Surg 1998; 133(1): 89-93.
[16] Haugen BR. Radioiodine remnant ablation: current indications and dosing regimens. Endocr Pract 2012; 18(4): 604-10.
[17] Lamartina L, Cooper DS. Radioiodine remnant ablation in low-risk differentiated thyroid cancer: the "con" point of view. Endocrine 2015; (In press).
[18] Antonelli A, Miccoli P, Ferdeghini M, *et al.* Role of neck ultra-sonography in the follow-up of patients operated on for thyroid cancer. Thyroid 1995; 5(1): 25-8.
[19] Antonelli A, Miccoli P, Fallahi P, *et al.* Role of neck ultra-sonography in the follow-up of children operated on for thyroid papillary cancer. Thyroid 2003; 13(5): 479-84.
[20] American Thyroid Association (ATA) Guidelines Taskforce on Thyroid Nodules and Differentiated Thyroid Cancer, Cooper DS, Doherty GM, Haugen BR, *et al.* Revised American Thyroid Association management guidelines for patients with thyroid nodules and differentiated thyroid cancer. Thyroid 2009; 19(11): 1167-214.
[21] Verburg FA, Mäder U, Tanase K, *et al.* Life expectancy is reduced in differentiated thyroid cancer patients ≥45 years old with extensive local tumor invasion, lateral lymph node, or distant metastases at diagnosis and normal in all other DTC patients. J Clin Endocrinol Metab 2013; 98(1): 172-80.
[22] Gruber JJ, Colevas AD. Differentiated thyroid cancer: focus on emerging treatments for radioactive iodine-refractory patients. Oncologist 2015; 20(2): 113-26.
[23] Antonelli A, Fallahi P, Ferrari SM, *et al.* Dedifferentiated thyroid cancer: A therapeutic challenge. Biomed Pharmacother 2008; 62(8): 559-63.
[24] Sun XS, Sun SR, Guevara N, *et al.* Indications of external beam radiation therapy in non-anaplastic thyroid cancer and impact of innovative radiation techniques. Crit Rev Oncol Hematol 2013; 86(1): 52-68.
[25] de Groot JW, Links TP, Plukker JT, Lips CJ, Hofstra RM. RET as a diagnostic and therapeutic target in sporadic and hereditary endocrine tumors. Endocr Rev 2006; 5(5): 535-60.
[26] Nikiforov YE. Thyroid carcinoma: molecular pathways and therapeutic targets. Mod Pathol 2008; 21(Suppl 2): S37-43
[27] Marotta V, Guerra A, Sapio MR, Vitale M. RET/PTC rearrangement in benign and malignant thyroid diseases: a clinical standpoint. Eur J Endocrinol 2011; 165(4): 499-507.
[28] Fagin JA. Challenging dogma in thyroid cancer molecular genetics- role of RET/PTC and BRAF in tumor initiation. J Clin Endocrinol Metab 2004; 89(9): 4264-6.
[29] Santoro M, Melillo RM, Fusco A. RET/PTC activation in papillary thyroid carcinoma: European Journal of Endocrinology Prize Lecture. Eur J Endocrinol 2006; 155(5): 645-53.
[30] Xing M. BRAF mutation in thyroid cancer. Endocr Relat Cancer 2005; 12(2): 245-62
[31] Liu X, Bishop J, Shan Y, *et al.* Highly prevalent TERT promoter mutations in aggressive thyroid cancers. Endocr Relat Cancer 2013; 20(4): 603-10.
[32] Xing M, Liu R, Liu X, *et al.* BRAF V600E and TERT promoter mutations cooperatively identify the most aggressive papillary thyroid cancer with highest recurrence. J Clin Oncol 2014; 32(25): 2718-26.
[33] Howell GM, Hodak SP, Yip L. RAS mutations in thyroid cancer. Oncologist 2013; 18(8): 926-32.
[34] Marotta V, Franzese MD, Del Prete M, *et al.* Targeted therapy with kinase inhibitors in aggressive endocrine tumors. Expert Opin Pharmacother 2013; 14(9): 1187-203.
[35] Tan A, Xia N, Gao F, Mo Z, Cao Y. Angiogenesis-inhibitors for metastatic thyroid cancer. Cochrane Database Syst Rev 2010; 3: CD007958.
[36] Bunone G, Vigneri P, Mariani L, *et al.* Expression of angiogenesis stimulators and inhibitors in human thyroid tumors and correlation with clinical pathological features. Am J Pathol 1999; 155(6): 1967-76.
[37] Cao Y. VEGF-targeted cancer therapeutics-paradoxical effects in endocrine organs. Nat Rev Endocrinol 2014; 10(9): 530-9.
[38] Tang C, Yang L, Wang N, *et al.* High expression of GPER1, EGFR and CXCR1 is associated with lymph node metastasis in papillary thyroid carcinoma. Int J Clin Exp Pathol 2014; 7(6): 3213-23.
[39] Lim DJ, Baek KH, Lee YS, *et al.* Clinical, histopathological, and molecular characteristics of papillary thyroid microcarcinoma. Thyroid 2007; 17(9): 883-8.

[40] Maxwell JE, Sherman SK, O'Dorisio TM, Howe JR. Medical management of metastatic medullary thyroid cancer. Cancer 2014; 120(21): 3287-301.
[41] Krampitz GW, Norton JA. RET gene mutations (genotype and phenotype) of multiple endocrine neoplasia type 2 and familial medullary thyroid carcinoma. Cancer 2014; 120(13): 1920-31.
[42] Antonelli A, Fallahi P, Ferrari SM, *et al.* RET TKI: potential role in thyroid cancers. Curr Oncol Rep 2012; 14(2): 97-104.
[43] Almeida MQ, Hoff AO. Recent advances in the molecular pathogenesis and targeted therapies of medullary thyroid carcinoma. Curr Opin Oncol 2012; 24(3): 229-34.
[44] Lyra J, Vinagre J, Batista R, *et al.* mTOR activation in medullary thyroid carcinoma with RAS mutation. Eur J Endocrinol 2014; 171(5): 633-40.
[45] Tamburrino A, Molinolo AA, Salerno P, *et al.* Activation of the mTOR pathway in primary medullary thyroid carcinoma and lymph node metastases. Clin Cancer Res 2012; 18(13): 3532-40.
[46] Lorusso PM, Eder JP. Therapeutic potential of novel selective-spectrum kinase inhibitors in oncology. Expert Opin Investig Drugs 2008; 17(7): 1013-28.
[47] Antonelli A, Fallahi P, Ferrari SM, *et al.* New targeted therapies for thyroid cancer. Curr Genomics 2011; 12(8): 626-31.
[48] Ye L, Santarpia L, Gagel RF. The evolving field of tyrosine kinase inhibitors in the treatment of endocrine tumors. Endocr Rev 2010; 31(4): 578-99.
[49] Jasim S, Ozsari L, Habra MA. Multikinase inhibitors use in differentiated thyroid carcinoma. Biologics 2014; 8: 281-91.
[50] Therasse P, Arbuck SG, Eisenhauer EA, *et al.* New guidelines to evaluate the response to treatment in solid tumors. European Organization for Research and Treatment of Cancer, National Cancer Institute of the United States, National Cancer Institute of Canada. J Natl Cancer Inst 2000; 92(3): 205-16.
[51] Baccarani M, Castagnetti F, Gugliotta G, Rosti G. A review of the European LeukemiaNet recommendations for the management of CML. Ann Hematol 2015; 94(Suppl 2): 141-7.
[52] Nishida T, Doi T, Naito Y. Tyrosine kinase inhibitors in the treatment of unresectable or metastatic gastrointestinal stromal tumors. Expert Opin Pharmacother 2014; 15(14): 1979-89.
[53] Fishman MN, Tomshine J, Fulp WJ, Foreman PK. A systematic review of the efficacy and safety experience reported for sorafenib in advanced renal cell carcinoma (RCC) in the post-approval setting. PLoS One 2015; 10(4): e0120877.
[54] Tejeda-Maldonado J, García-Juárez I, Aguirre-Valadez J, *et al.* Diagnosis and treatment of hepatocellular carcinoma: An update. World J Hepatol 2015; 7(3): 362-76.
[55] Fallahi P, Ferrari SM, Santini F, *et al.* Sorafenib and thyroid cancer. BioDrugs 2013; 27(6): 615-28.
[56] Chougnet CN, Schlumberger M, Leboulleux S, Baudin E. Vandetanib, in the management of patients with locally advanced or metastatic medullary thyroid carcinomas. Bull Cancer 2014; 101(9): 891-5.
[57] Durante C, Paciaroni A, Plasmati K, Trulli F, Filetti S. Vandetanib: opening a new treatment practice in advanced medullary thyroid carcinoma. Endocrine 2013; 44(2): 334-42.
[58] Nagilla M, Brown RL, Cohen EE. Cabozantinib for the treatment of advanced medullary thyroid cancer. Adv Ther 2012; 29(11): 925-34.
[59] Goldenberg MM. Pharmaceutical approval update. P T 2013; 38(2): 86-95.
[60] van Leeuwen RW, van Gelder T, Mathijssen RH, Jansman FG. Drug-drug interactions with tyrosine-kinase inhibitors: a clinical perspective. Lancet Oncol 2014; 15(8): e315-26.
[61] Ruan M, Liu M, Dong Q, Chen L. Iodide- and glucose-handling gene expression regulated by sorafenib or cabozantinib in papillary thyroid cancer. J Clin Endocrinol Metab 2015; 100(5): 1771-9.
[62] Gupta-Abramson V, Troxel AB, Nellore A, *et al.* Phase II trial of sorafenib in advanced thyroid cancer. J Clin Oncol 2008; 26(29): 4714-9.
[63] Kloos RT, Ringel MD, Knopp MV, *et al.* Phase II trial of sorafenib in metastatic thyroid cancer. J Clin Oncol 2009; 27(10): 1675-84.
[64] Keefe SM, Troxel AB, Rhee S, *et al.* Phase II trial of sorafenib in patients with advanced thyroid cancer. ASCO Annual Meeting, Pennsylvania, Philadelphia, 2011. J Clin Oncol 2011; 29 (suppl 5562).
[65] Schneider TC, Abdulrahman RM, Corssmit EP, Morreau H, Smit JW, Kapiteijn E. Long-term analysis of the efficacy and tolerability of sorafenib in advanced radio-iodine refractory differentiated thyroid carcinoma: final results of a phase II trial. Eur J Endocrinol 2012; 167(5): 643-50.

[66] Marotta V, Ramando V, Camera L, *et al.* Sorafenib in advanced iodine-refractory differentiated thyroid cancer: efficacy, safety and exploratory analysis of role of serum thyroglobulin and FDG-PET. Clin Endocrinol (Oxf) 2013; 78(5): 760-7.
[67] Massicotte MH, Brassard M, Claude-Desroches M, *et al.* Tyrosine kinase inhibitor treatments in patients with metastatic thyroid carcinomas: a retrospective study of the TUTHYREF network. Eur J Endocrinol 2014; 170(4): 575-82.
[68] Brose MS, Nutting CM, Jarzab B, *et al.* Sorafenib in radioactive iodine-refractory, locally advanced or metastatic differentiated thyroid cancer: a randomised, double-blind, phase 3 trial. Lancet 2014; 384(9940): 319-28.
[69] Bayer. Nexavar® *versus* placebo in locally advanced/metastatic RAI-refractory differentiated thyroid cancer [ClinicalTrials.gov identifier: NCT00984282] (2015). Available from: https://clinicaltrials.gov/ct2/show/NCT00984282?term=NCT00984282&rank=1
[70] Frampton JE. Vandetanib: in medullary thyroid cancer. Drugs 2012; 72(10): 1423-36.
[71] Carlomagno F, Vitagliano D, Guida T, *et al.* ZD6474, an orally available inhibitor of KDR tyrosine kinase activity, efficiently blocks oncogenic RET kinases. Cancer Research 2002; 62(24): 7284-90.
[72] Carlomagno F, Guida T, Anaganti S, *et al.* Disease associated mutations at valine 804 in the RET receptor tyrosine kinase confer resistance to selective kinase inhibitors. *Oncogene* 2004; 23(36): 6056-63.
[73] Johanson V, Ahlman H, Bernhardt P, *et al.* A transplantable human medullary thyroid carcinoma as a model for RET tyrosine kinase-driven tumorigenesis. Endocr Relat Cancer 2007; 14(2): 433-44.
[74] Ryan AJ, Wedge SR. ZD6474: a novel inhibitor of VEGFR and EGFR tyrosinekinase activity. Br J Cancer 2005; 92(Suppl 1): 6-13.
[75] Morabito A, Piccirillo MC, Falasconi F, *et al.* Vandetanib (ZD6474), a dual inhibitor of vascular endothelial growth factor receptor (VEGFR) and epidermal growth factor receptor (EGFR) tyrosine kinases: current status and future directions. Oncologist 2009;14(4): 378-90.
[76] Wells SA Jr, Robinson BG, Gagel RF, *et al.* Vandetanib in patients with locally advanced or metastatic medullary thyroid cancer: a randomized, double-blind phase III trial. J Clin Oncol 2012; 30(2): 134-41.
[77] Cooper MR, Yi SY, Alghamdi W, Shaheen DJ, Steinberg M. Vandetanib for the treatment of medullary thyroid carcinoma. Ann Pharmacother 2014; 48(3): 387-94.
[78] Chougnet CN, Borget I, Leboulleux S, *et al.* Vandetanib for the treatment of advanced medullary thyroid cancer outside a clinical trial: results from a French cohort. Thyroid 2015; 25(4): 386-91.
[79] AstraZeneca to compare the effects of two doses of vandetanib in patients with advanced medullary thyroid cancer. [Accessed February 27, 2015]. Available from: https://www.clinicaltrials.gov/ct2/show/NCT01496313.
[80] AstraZeneca evaluation of efficacy, safety of vandetanib in patients with differentiated thyroid cancer (VERIFY). [Accessed February 27, 2015]. Available from: https://www.clinicaltrials.gov/ct2/show/NCT01876784.
[81] Yakes FM, Chen J, Tan J, *et al.* Cabozantinib (XL184), a novel MET and VEGFR2 inhibitor, simultaneously suppresses metastasis, angiogenesis, and tumor growth. Mol Cancer Ther 2011; 10(12): 2298-308.
[82] Grüllich C. Cabozantinib: a MET, RET, and VEGFR2 tyrosine kinase inhibitor. Recent Results Cancer Res 2014; 201: 207-14.
[83] Niafar M, Dabiri S, Bozorgi F, Niafar F, Gholami N. Metastatic medullary thyroid carcinoma: A case report. J Res Med Sci 2011; 16(4): 568-73.
[84] Viola D, Cappagli V, Elisei R. Cabozantinib (XL184) for the treatment of locally advanced or metastatic progressive medullary thyroid cancer. Future Oncol 2013; 9(8): 1083-92.
[85] Elisei R, Schlumberger MJ, Müller SP, *et al.* Cabozantinib in progressive medullary thyroid cancer. J Clin Oncol 2013; 31(29): 3639-46.
[86] Decillis A. Cabozantinib dosage form and use in the treatment of cancer. WO 2014165786, 2013.
[87] Hoy SM. Cabozantinib: a review of its use in patients with medullary thyroid cancer. Drugs 2014; 74(12): 1435-44.
[88] Wang L, Wang X. Cabozantinib capsule composition and preparation method thereof. CN 103830203 A, 2014.

[89] Peng Z, Li R. Cabozantinib dispersible tablets and preparation method thereof. CN 103751140 A, 2014.

[90] Matsui J, Yamamoto Y, Funahashi Y, *et al.* E7080, a novel inhibitor that targets multiple kinases, has potent antitumor activities against stem cell factor producing human small cell lung cancer H146, based on angiogenesis inhibition. Int J Cancer 2008; 122(3): 664-71.

[91] Matsui J, Funahashi Y, Uenaka T, Watanabe T, Tsuruoka A, Asada M. Multikinase inhibitor E7080 suppresses lymphnode and lung metastases of human mammary breast tumor MDA-MB-231 *via* inhibition of vascular endothelial growth factor-receptor (VEGF-R) 2 and VEGF-R3 kinase. Clin Cancer Res 2008; 14(17): 5459-65.

[92] Sherman SI, Jarzab B, Cabanillas ME, *et al.* A phase II trial of the multitargeted kinase inhibitor E7080 in advanced radioiodine (RAI)-refractory differentiated thyroid cancer (DTC) [abstract]. J Clin Oncol 2011; 29(Suppl 15): 5503.

[93] Schlumberger M, Tahara M, Wirth LJ, *et al.* Lenvatinib *versus* placebo in radioiodine-refractory thyroid cancer. N Engl J Med 2015; 372(7): 621-30

[94] Antonelli A, Bocci G, La Motta C, *et al.* Novel pyrazolopyrimidine derivatives as tyrosine kinase inhibitors with antitumoral activity *in vitro* and *in vivo* in papillary dedifferentiated thyroid cancer. J Clin Endocrinol Metab 2011; 96(2): E288-96.

[95] Bocci G, Fioravanti A, La Motta C, *et al.* Antiproliferative and proapoptotic activity of CLM3, a novel multiple tyrosine kinase inhibitor, alone and in combination with SN-38 on endothelial and cancer cells. Biochem. Pharmacol 2011; 81(11): 1309-16.

[96] Sartini S, Coviello V, Bruno A, *et al.* Structure-based optimization of tyrosine kinase inhibitor CLM3. Design, synthesis, functional evaluation, and molecular modeling studies. J Med Chem 2014; 57(54): 1225-35.

[97] Antonelli A, Bocci G, Fallahi P, *et al.* CLM3, a multitarget tyrosine kinase inhibitor with antiangiogenic properties, is active against primary anaplastic thyroid cancer *in vitro* and *in vivo*. J Clin Endocrinol Metab 2014; 99(4): E572-81.

[98] Antonelli A, Bocci G, La Motta C, *et al.* CLM94, a novel cyclic amide with anti-VEGFR-2 and antiangiogenic properties, is active against primary anaplastic thyroid cancer *in vitro* and *in vivo*. J Clin Endocrinol Metab 2012; 97(4): E528-36.

[99] Tan K, Lawler J. The interaction of Thrombospondins with extracellular matrix proteins. J Cell Commun Signal 2009; 3(3-4): 177-87.

[100] Zhang X, Kazerounian S, Duquette M, *et al.* Thrombospondin-1 modulates vascular endothelial growth factor activity at the receptor level. FASEB J 2009; 23(10): 3368-76.

[101] Ferrari SM, Fallahi P, La Motta C, *et al.* Antineoplastic activity of the multitarget tyrosine kinase inhibitors CLM3 and CLM94 in medullary thyroid cancer *in vitro*. Surgery 2014; 156(5): 1167-76.

[102] Antonelli A, Bocci G, La Motta C, *et al.* CLM29, a multi-target pyrazolopyrimidine derivative, has anti-neoplastic activity in medullary thyroid cancer *in vitro* and *in vivo*. Mol Cell Endocrinol 2014; 393(1-2): 56-64.

CHAPTER 4

Shaping the Genome with Non-coding RNAs

Xue Qing David Wang, Jennifer Lynn Crutchley and Josée Dostie*

Department of Biochemistry, and Goodman Cancer Research Center, McGill University, 3655 Promenade Sir-William-Osler, Room 815A, Montréal, Québec, H3G1Y6, Canada

Abstract: The human genome must be tightly packaged in order to fit inside the nucleus of a cell. Genome organization is functional rather than random, which allows for the proper execution of gene expression programs and other biological processes. Recently, three-dimensional chromatin organization has emerged as an important transcriptional control mechanism. For example, enhancers were shown to regulate target genes by physically interacting with them regardless of their linear distance and even if located on different chromosomes. These chromatin contacts can be measured with the "chromosome conformation capture" (3C) technology and other 3C-related techniques. Given the recent innovation of 3C-derived approaches, it is not surprising that we still know very little about the structure of our genome at high-resolution. Even less well understood is whether there exist distinct types of chromatin contacts and importantly, what regulates them. A new form of regulation involving the expression of long non-coding RNAs (lncRNAs) was recently identified. lncRNAs are a very abundant class of non-coding RNAs that are often expressed in a tissue-specific manner. Although their different subcellular localizations points to their involvement in numerous cellular processes, it is clear that lncRNAs play an important role in regulating gene expression. They have been shown to bind several ribonucleoprotein complexes including the polycomb repression complex, which modify chromatin epigenetically by transferring histone marks, but how they control transcription however is mostly unknown. In this review, we provide an overview of known lncRNA transcription regulation activities. We also discuss potential mechanisms by which ncRNAs might exert three-dimensional transcriptional control and what recent studies have revealed about their role in shaping our genome.

Keywords: Chromatin, chromosome conformation capture, epigenetics, gene regulation, histone marks, non-coding RNA, polycomb repression complex, ribonucleoprotein complex, three-dimensional chromatin organization, transcription.

INTRODUCTION

Organisms with large genomes face the interesting problem of having to contain their genetic material within very small cellular volumes. For example, a human

***Corresponding author Josée Dostie:** Department of Biochemistry, and Goodman Cancer Research Center, McGill University, 3655 Promenade Sir-William-Osler, Room 815A, Montréal, Québec, H3G1Y6, Canada; Tel: 1-(514)-398-4975; Fax: 1-(514)-398-7384; E-mail: josee.dostie@mcgill.ca

Atta-ur-Rahman (Ed.)

diploid genome measures approximately 2 meters in length and must fit within a micron-sized nucleus. Our genomic DNA must consequently be extensively packaged in a functional manner that allows activities like transcription, DNA replication and repair to occur properly. Although exactly how our genome is functionally organized *in vivo* is poorly understood, several key features appear common to most cell types (Fig. **1**). First, chromosomes are known to occupy distinct nuclear areas that are termed "chromosome territories" (CT). Gene-rich chromosomes tend to localize to the center of the nucleus and gene-poor near the periphery (reviewed in [1]). Genomic domains with similar activities and co-regulated genes often co-localize *in vivo* (reviewed in [2]). Moreover, genes located near anchor points where interphase chromosomes attach to the nuclear matrix tend to be poorly transcribed (reviewed in [3]). Together, these observations clearly reveal the existence of an intimate relationship between the organization of our genome in the nuclear space and gene activity.

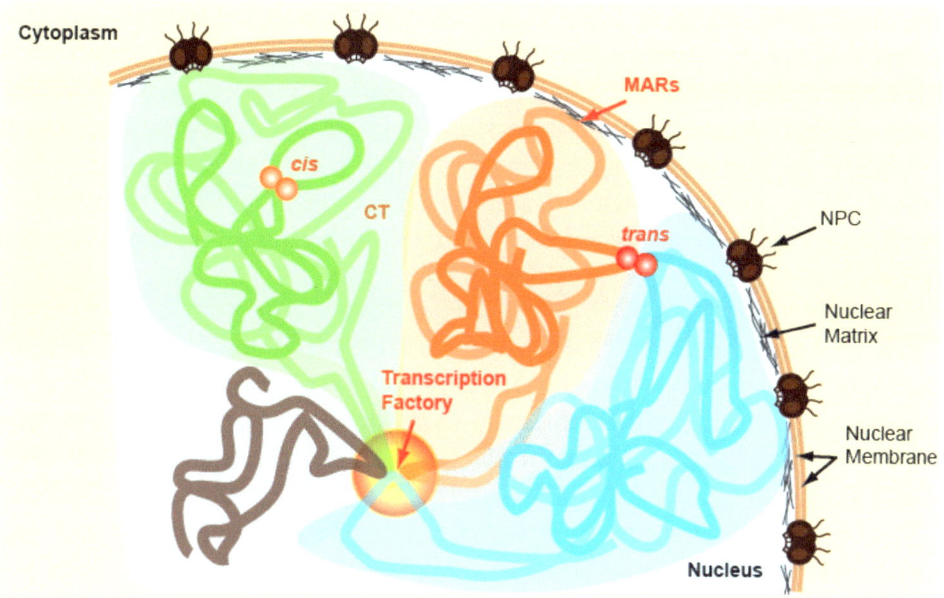

Fig. (1). *In vivo genome organization.* Individual chromosomes are illustrated by thick green, orange, blue and grey strokes. Respective chromosome territories (CT) are highlighted in corresponding colors. Chromatin interactions within (*cis*) and between (*trans*) chromosomes are represented by orange and red spheres. Transcription factory is shown as a graded circle. MARs: matrix attachment regions. NPC: nuclear pore complex.

At the molecular level, spatial genome organization can critically affect gene expression. *In vivo*, genomic DNA exists in the form of chromatin where DNA is associated with histone and non-histone proteins. While it is well known that chromatin composition can directly shape gene activity [4-6], three-dimensional chromatin organization is also emerging as an important gene regulation mechanism. For example, it was shown that distal control DNA elements could regulate target genes located on the same or on different chromosomes by physically interacting with them [7-9]. This type of long-range regulation has been reported for genes located throughout the genome that are involved in various cellular pathways [10-14]. Consequently, the genome is now viewed as a functional network of physical contacts within (*cis*) and between (*trans*) chromosomes. These physical DNA contacts can be mapped at high-resolution with technologies such as the "chromosome conformation capture" (3C) and other 3C-related methods including 3C-Carbon Copy (5C) and Hi-C [15-20]. Given the recent development and sophistication of these approaches, it is not surprising that we still know very little about genome organization at the molecular level. In fact, most questions remain unanswered: How many types of contacts exist? What are their relative contributions to overall genome organization and to various genome functions? What are their tissue-specificities? What are their roles in the establishment and/or maintenance of cellular identity? What is their involvement in human disease? Answers to these questions will require extensive mapping and classification of chromatin interactions genome-wide under various cellular conditions. Only then will we be able to distinguish between regulated and structural contacts, and understand the outcome of altering either type on cellular function.

Even less well understood is what mediates chromatin interactions and how they are regulated. It was shown that chromatin contacts could be mediated by tissue-specific transcription factors. Such is the case for the beta-globin cluster where DNA looping between the locus control region (LCR) and activated beta-globin genes was shown to require the GATA-1 transcription factor and co-factor FOG-1 [21]. This type of long-range control mechanism where enhancers and distal promoters are brought in close physical proximity to each other by transcription factors has been found at other loci. These include many estrogen-regulated genes [22], the B lymphocyte-specific induction of *CIITA* by PU.1 [23], and *MYC* activation by TCF and the beta-catenin co-activator complex [24].

Other chromatin-binding proteins such as the CCCTC-binding factor (CTCF) and cohesin also appear to play critical roles in genome organization and gene expression [25-28]. Mammalian CTCF is a DNA-binding protein associated with insulator sequences, boundary elements and imprinting control regions, all of which are thought to organize our genome into functional subdomains. CTCF is known to form networks of chromatin loops genome-wide and is viewed as a master regulator of spatial genome organization. It was also shown that CTCF co-localizes extensively with cohesin during interphase by directly binding to its STAG1/2 component [29]. Cohesin is a four protein subunit ring complex important in DNA repair, chromosome segregation and transcription regulation. Within some genomic regions, CTCF-co-localized cohesin rather than solely CTCF itself appears to be responsible for chromatin contacts. For example, cohesin knockdown was shown to abolish long-range *cis* chromatin contacts associated with *IFNG* locus expression in T helper 1 cells, and to reduce gene expression without affecting CTCF binding [30]. Similar cohesin-dependent loops and gene expression effects have also been observed at other loci. These include the beta-globin locus [31] and the apolipoprotein gene cluster [32]. Although they are often found together on the chromatin, CTCF and cohesin do not however exclusively co-localize with each other. Since cohesin does not bind DNA directly, this observation points to the existence of CTCF-independent cohesin recruitment and looping mechanisms. One such mechanism could involve tissue-specific transcription factors. Indeed, CTCF-independent cohesin binding sites frequently co-localize with tissue-specific transcription factors [33]. Alternatively, transcriptional coactivators such as mediator could recruit cohesin [34].

There are many ways by which chromatin interactions could be regulated. First, by modifying the DNA itself with cytosine methylation and consequently altering protein association. An example for this type of regulation is found at the imprinted *Igf2/H19* locus. On the maternal allele, expression of the *Igf2* gene is inhibited through the formation of a repressive loop encompassing the gene. This loop requires the recruitment of cohesin through CTCF bound at the DMR1 (DNA methylated region 1) and ICR (imprinted control region) sequences located upstream of *Igf2* and *H19* respectively [35]. Bound CTCF was also shown to recruit the polycomb repressive complex 2 (PRC2) at the *Igf2* promoter through its Suz12 subunit, thereby leading to allele-specific trimethylation of histone H3 lysine 27 (H3K27me3) and suppression of the maternal *Igf2* allele [36]. In contrast, binding of CTCF to the ICR is negatively regulated by DNA methylation on the paternal allele. This lack of binding in turn prevents formation of the repressive loop, and enables access of the *Igf2* promoter to an enhancer located downstream of *H19* [37, 38].

Chromatin contacts could also be regulated by controlling access to DNA sequences with post-translational histone modifications (PTMs), the use of histone variants or by altering nucleosome positioning. Similarly, post-translational modification or changes in expression level of non-histone chromatin-binding proteins could represent important mechanisms to regulate chromatin contacts. Additionally, non-coding RNAs (ncRNAs) and their protein complexes could regulate the three-dimensional architecture of our genome. ncRNAs are a broad class of transcripts consisting of structural (rRNAs, tRNAs, snRNAs, snoRNAs, *etc.*), regulatory (miRNAs, piRNAs, *etc.*), and of sense/anti-sense transcripts, whose functions remain mostly uncharacterized. The latter RNA class includes transcriptional "features" (eRNAs, tiRNAs), and a very large number of long non-coding RNAs (lncRNAs) ranging in length from 200 nt to 100 kb. Although lncRNAs can be transcribed nearby known protein-coding genes or from their introns, they are often produced intergenically and hence, are sometimes referred to as "lincRNAs". While a number of transcripts falling in that category have been studied for some time (*e.g.* Xist, see below), the realization that lncRNAs represent a very abundant RNA subclass is relatively recent and originates from genome-wide transcriptome studies [39-46]. lncRNAs are estimated to be at least as abundant as mRNAs, and like mRNAs, their expression levels and patterns are often tissue-specific. The heterogeneity in sequence, structure and size, along with the widely different subcellular localizations of lncRNAs points to their involvement in numerous cellular processes [47]. In the nucleus, a few lncRNAs were shown to inhibit or activate transcription, which is in contrast to other ncRNAs thought to represent basic features of transcription.

Until recently, DNA sequence and associated proteins were thought to mainly dictate spatial genome organization. With the abundance and diversity of lncRNAs rivaling that of mRNAs, it seems likely that they also play an essential role in regulating chromatin organization. In fact, there is already evidence for the implication of ncRNA in general genome architecture [48], and in maintaining the integrity of nuclear bodies [49]. Conversely, chromatin conformation is thought to play a role in coordinating gene expression with lncRNAs [50]. Below, we report on the nuclear activity of a number of best-characterized lncRNAs, and discuss how lncRNAs might be instrumental in shaping the three-dimensional organization of our genome.

lncRNAs as Inhibitors of Transcription

Although most lncRNAs have only recently been discovered, a few transcripts belonging to this category have already been scrutinized for some time. Amongst these, the X-inactive specific transcript (Xist) responsible for equalizing gene expression between mammalian males and females was first identified in the early

90's [51, 52]. As described below, this lncRNA silences genes in *cis* by coating the chromosome. More recently, a lncRNA termed HOTAIR was shown to function in *trans*. Like the Xist transcript, HOTAIR associates with the polycomb repressive complex 2 (PRC2) and is thought to repress transcription by increasing H3K27me3 levels at specific genes. How these transcripts bind and help target polycomb complexes specifically to specific regions however is unknown. lncRNAs are also important for gene imprinting. For example, the Air lncRNA was shown to silence 3 genes in *cis* on the paternal allele at the *Kcnq-1* imprinted region. Like Xist, the Air transcript coats adjacent chromatin, but instead recruits the G9a histone 3 lysine 9 methyltransferase (H3K9me3) to repress transcription. Interestingly, other lncRNAs such as Kcnq1ot1 silence by binding both PRC2 and G9a to increase the levels of H3K9me3 and H3K27me3. Although repression by PRC2 binding and elevated H3K27me3 is thus far the most common mode of inhibition (see also ANRIL below), we can expect that many types of mechanisms will mediate lncRNA activity in *cis* and *trans* (Fig. **2**). In fact, alternative mechanisms have already been identified. For example, the p53-inducible lincRNA-p21 was shown to bind hnRNP-K and inhibit gene expression genome-wide following p53 induction. We report below on the function and mechanism of some of these inhibitory lncRNAs.

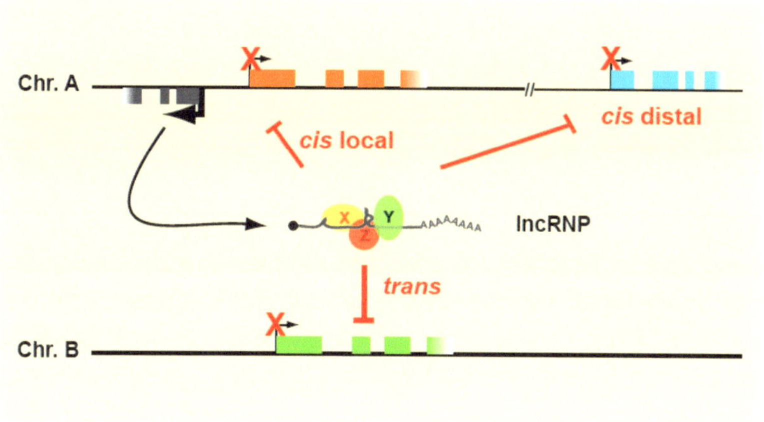

Fig. (2). *lncRNAs as inhibitors of transcription.* lncRNAs expressed from anywhere in the genome can be packaged with transcript-specific proteins into functional lncRNPs (long non-coding ribonucleoproteins). The X, Y and Z ovals represent putative transcript-specific proteins associated with lncRNAs. lncRNPs could act locally or distally in *cis*, or in *trans*. Protein-coding genes are represented by orange, blue and green boxes. ncRNA gene is illustrated in grey. Transcription start sites (TSSs) are indicated by bent arrows, with thick arrows showing greater transcription activity. TSS of inhibited genes are indicated by thin bent arrows marked with an "X".

Tsix Repression of Xist

In mammals, flies and worms, females differ from males by having two X chromosomes rather than one. The abundance of X-linked gene products must therefore be equalized between sexes in the somatic cells of these organisms. In mammals, gene dosage compensation is achieved by inactivating one X chromosome in females through a process called X chromosome inactivation (XCI). XCI requires the Xist lncRNA expressed in *cis*, which coats one female X chromosome and inactivates it by recruiting polycomb proteins (reviewed in [53, 54]). Xist is expressed from a region on the X chromosome called X inactivation center (*Xic*) where its transcription is controlled by multiple lncRNAs. Indeed, the *Xic* encodes at least 6 other lncRNAs, 2 of which are known to antagonistically regulate Xist expression. On the one hand, the Tsix lncRNA transcribed antisense to *Xist* is known to repress Xist expression on the active X (Xa). Conversely, the Jpx lncRNA activates *Xist* on the inactive X (Xi). Tsix represses *Xist* transcription *via* multiple mechanisms that include recruiting DNA methyltransferase activity and the RNA interference machinery locally. Tsix is also thought to change the epigenetic chromatin state of *Xist*, thereby preventing Xist and RepA (a short transcript corresponding to the 5'end of Xist) from recruiting PRC2 (reviewed in [53]). Interestingly, the *Tsix* gene itself is positively regulated on Xa by another *Xic*-encoded lncRNAs called Xite [55]. How Jpx regulates *Xist* will be described below in the next section. Together, XCI studies reveal that lncRNAs can regulate the transcription of other ncRNAs in addition to protein-coding genes. lncRNAs may therefore form very complex regulatory networks that control transcription genome-wide.

HOTAIR

In 2007, Chang and colleagues reported the first comprehensive analysis of the transcriptional landscapes of the four human *HOX* clusters in 11 different fibroblast cell types [56]. In this study, 231 Hox ncRNAs were identified at 5-bp resolution on custom tilling arrays. Like *HOX* genes, these ncRNAs displayed differential expression along developmental axes and marked chromosomal domains with different histone methylation patterns and RNA pol II accessibility. The *HOX* clusters encode highly conserved eukaryotic transcription factors with pivotal roles in body patterning during development. *HOX* genes were originally identified in *Drosophila melanogaster* and their mammalian counterparts are known to have similar developmental functions. For example, studies in mouse models demonstrated that dysregulated *HOX* gene expression in early development could lead to severe limb and skeletal malformations [57-59]. *HOX* gene expression must also be controlled in adult tissues as overexpression of some *HOX* genes was found to be a hallmark of certain cancers such as leukemia [60, 61]. Although *HOX* regulation is clearly essential during development and for human health, how these genes are controlled remains poorly understood to this day. *HOX* clusters are regulated epigenetically and in pluripotent cells rest in so-called bi-valent domains where activating H3K4me3 and repressive H3K27me3 histone marks co-exist [62, 63]. The study led by Chang and colleagues uncovered not only a promising new level of *HOX* regulation, but also showed for the first time that lncRNAs can regulate genes in *trans*. Indeed, a lncRNA transcribed at a boundary between active and repressive chromatin states in the *HOXC* cluster was found to repress the expression of 5' end *HOXD* genes located on another chromosome. This RNA is transcribed in an anti-sense manner to the *HOXC* genes and was named HOTAIR for "*HOX* Antisense Intergenic RNA". HOTAIR is a 2,158 nt long spliced and polyadenylated RNA that binds PRC2 and silences genes by increasing H3K27me3 levels at specific sites [56, 64]. While the 5' end of HOTAIR was shown to bind PRC2 directly through its Suz12 subunit, its 3' end was also shown to bind the LSD1/CoREST/REST H3K4 demethylase complex. Thus, HOTAIR is thought to provide a scaffold for enzyme complexes that increase H3K27me3 (PRC2) and decrease H3K4me3 levels (LSD1/coREST/REST) to silence gene expression [64].

In addition to the *HOX* genes themselves, which are involved in certain types of leukemia, Hox ncRNAs were also shown to be dysregulated in breast cancer [65]. Amongst these, high HOTAIR expression was found to correlate well with increased metastasis and lower patient life expectancy. Conversely, loss of HOTAIR expression reduced cancer invasiveness suggesting that lncRNAs, particularly HOTAIR, might significantly alter the epigenome of cancer cells. The association of lncRNAs with polycomb complexes sparks numerous questions: How can so many different lncRNA sequences bind directly to PRC2? How do these transcripts help target protein complexes to specific genes? Does the actual lncRNA sequence actively participate in targeting the silencing complexes to given genes? Might the local accumulation of lncRNAs around their transcription sites be sufficient to recruit histone-modifying enzymes and alter chromatin states?

Local INK4/ARF Locus Control by the ANRIL Non-Coding Transcript

The human *INK4/ARF* locus encodes three tumor suppressor genes named $p14^{ARF}$, $p15^{INK4B}$, and $p16^{INK4A}$ that are often altered in cancers [66]. In normal tissues, these genes are epigenetically repressed by PRC2-mediated deposition of

H3K27me3, and are activated upon cellular stress to halt proliferation by inhibiting CDK4, CDK6 and MDM2. In 2007, a lncRNA transcribed anti-sense to the three tumor suppressor genes was identified along the *INK4/ARF* locus in human testes [67]. The mature 3,834-nt transcript was named ANRIL for "Antisense Non-coding RNA in the *INK4* Locus" and was derived from a 126 kb gene composed of 19 exons. In addition to full-length ANRIL, a shorter 2,659-nt form was also found in the same tissue, and 8 additional splice variants were identified in several cell lines [68]. Although the roles of individual forms are unknown, full-length ANRIL was shown to repress $p15^{INK4B}$ expression [69]. Indeed, ANRIL knockdown in actively growing cells significantly increased expression of $p15^{INK4B}$. Similarly, induction of $p15^{INK4B}$ and $p16^{INK4A}$ by oncogenic Ras was found to repress ANRIL expression. It was known that the three genes were silenced by PRC2 under normal cellular condition, and ANRIL knockdown significantly reduced PRC2 occupancy and H3K27me3 at the *INK4/ARF* locus. ANRIL was also found to co-immunoprecipitate with Suz12 and CBX7, which are components of PRC2 and the polycomb repressive complex 1 (PRC1), respectively [69, 70]. Like PRC2, PRC1 is a multisubunit complex and can bind H3K27me3 through its CBX components [71]. PRC1 is also responsible for the ubiquitination of H2AK119 known to inhibit RNA pol II transcription elongation [72-75]. Thus, the lncRNA ANRIL silences the $p15^{INK4B}$ gene by recruiting both polycomb repressive complexes to halt RNA pol II transcription elongation.

Another anti-sense transcript named p15-antisense (p15AS) was identified at the *INK4/ARF* locus in 2008 [76]. Unlike ANRIL, this smaller RNA measures 34.8 kb in length and overlaps only the $p15^{INK4B}$ gene. Consistent with the frequent epigenetic silencing of $p15^{INK4B}$ in leukemia patient samples, higher p15AS levels were often found in acute lymphoblastic (ALL) and myeloid leukemias (AML). Thus, it was suggested that p15AS could facilitate cancer progression by silencing the $p15^{INK4B}$ tumor suppressor gene. Surprisingly, p15AS overexpression increased the repressive H3K9me2 mark rather than H3K27me3, and decreased the levels of activating H3K4me2 near the TSS of the endogenous and exogenous $p15^{INK4B}$ gene. Similar histone modifications were also found at the $p15^{INK4B}$ promoter in cells where the gene is silent and no DNA methylation changes were observed. These results indicate that p15AS is capable of regulating its parental gene in *cis* and in *trans* through the formation of a type of heterochromatin. Although the silencing mechanism has not yet been identified, p15AS appears to trigger silencing but not maintain it since removal of the transcript did not reverse its effects.

Gene Imprinting by Air and Kcnq1ot1

RNA transcripts important for gene imprinting were amongst the first lncRNAs identified in human and mouse. Gene imprinting is an epigenetic process that selectively represses one or more genes from a parental allele. This process was originally thought to result mainly from differential DNA methylation at CpG islands [77], but there is now overwhelming evidence that histone-modifying complexes and lncRNAs also play a significant role in this process. The Air and Kcnq1ot1 transcripts are two well-studied lncRNAs involved in gene imprinting [78, 79]. Interestingly, these transcripts differ from Xist, HOTAIR and ANRIL by their much larger sizes and absence of splicing [79, 80]. The Air transcript (antisense *Igf2r* RNA) is a 107,796-nt long antisense RNA transcribed from the second intron of the *Igf2r* gene to the 3' end of the *Mas* gene [81]. Air is expressed only from the paternal allele where it silences the overlapping *Igf2r* gene and two other genes (*Slc22a2 and Slc22a3*) downstream of *Igf2r* [82]. The three genes are not consecutive and span approximately 300 kb indicating that Air acts specifically in *cis*. Although it is not spliced, Air is transcribed by RNA pol II and thus bears a 5' m(7)GpppN cap structure [83]. The full-length RNA transcript appears to be required for imprinting since deletion of its promoter or gene truncation prevented paternal silencing of the three genes [79]. Similarly to Xist, the Air lncRNA was found to coat its target genes but instead recruits the G9a histone methyltransferase rather than the subunits of PRC2 [84]. G9a is a SET domain-containing histone methyltransferase similar to polycomb Ezh2 but preferentially methylates H3 on lysine 9 (H3K9me3) [85-87]. Truncation substantially decreased Air lncRNA concentration around target genes, reduced G9a recruitment at the *Igf2r* locus, and failed to inhibit the *Igf2r*, *Slc22a2 and Slc22a3* genes [84].

The kcnq1ot1 transcript is another well-studied lncRNA involved in imprinting [88, 89]. Kcnq1ot1 (also called lit1 for "long QT intronic transcript 1") is an antisense transcript that shares many common attributes with Air. First, it is a 91 kb RNA transcribed by RNA pol II, is capped, polyadenylated, but not spliced [80]. The transcript is expressed from an unmethylated promoter on the paternal allele of the *kcnq1* (potassium voltage-gated channel, KQT-like subfamily, member 1) gene and stays in the nucleus to silence protein coding genes in *cis* within the region [88, 90, 91]. In contrast to Air, which represses three genes at the *Igf2r* locus, Kcnq1ot1 can silence up to 10 genes in the *Kcnq1* cluster. Surprisingly, Kcnq1ot1 interacts with both G9a and the Ezh2 and Suz12 components of the PRC2 complex, and silences the surrounding genes by spreading H3K9me3 and H3K27me3 marks [80, 92]. A silencing domain was also identified at the 5' end of the Kcnq1ot1 and further analysis revealed that Kcnq1ot1 silences gene expression not only by recruiting

HMTs to remodel chromatin structure, but also by re-localizing its target genes near heterochromatin at the perinucleolar region [93].

LincRNA-p21

The *p53* tumor suppressor gene encodes a key transcription factor, which is mutated in over 50% of human cancers. p53 is induced by many types of cellular stresses and exerts its tumor suppressor properties primarily through its ability to act as a sequence-specific transcription factor [94-97]. Many genes are upregulated or downregulated during the p53 transcriptional response. While p53-mediated gene activation is relatively well understood, how this transcription factor could lead to gene repression was unknown until recently. As lncRNAs had been shown to silence genes by recruiting chromatin-modifying complexes, it seemed possible that p53 might repress genes by activating lncRNAs. In 2010, Rinn and colleagues used high-resolution tilling arrays featuring predicted lncRNAs to probe the ncRNA transcriptome of p53-induced cells [98]. The custom tilling arrays included 400 genomic regions outside known protein-coding genes that were selected based on the presence of active H3K4me3 and H3K36me3 histone marks [99, 100]. Over 30 lncRNAs were induced by p53 in two different cellular systems indicating that like protein-coding genes, numerous lncRNAs are also temporally regulated during the p53 transcriptional response. Two lncRNAs containing canonical p53 binding sites in their promoters were shown to be *bona fide* p53 transcription targets and one was retained for further analysis. The lncRNA named lincRNA-p21 was interesting because of its curious genomic position approximately 15 kb upstream of the p21 cell cycle regulator, which is a well-known p53 target gene. lincRNA-p21 is a 3.1 kb spliced and polyadenylated RNA transcribed from the opposite strand to *p21*, and is conserved in human and mouse. Individual knockdown of p53 and lincRNA-p21 revealed a strong overlap in derepressed genes indicating that lincRNA-p21 participates in p53 gene repression. Common derepressed genes were enriched in those involved in cell cycle arrest and apoptosis although further analysis indicated that lincRNA-p21 increases apoptosis without any significant effects on the cell cycle. Intriguingly, lincRNA-p21 was found to interact with hnRNP-K through a 780 nt region at its 5'end. Although deletion of this region abolished the ability of lincRNA-p21 to induce apoptosis, the region alone was not sufficient to trigger it. Furthermore hnRNP-K knockdown also showed a strong overlap in derepressed genes with lincRNA-p21 and p53, indicating that hnRNP-K is involved in the repression of many p53 downregulated genes likely through lincRNA-p21. Accordingly, the authors found that hnRNP-K is at the promoter of many p53-repressed genes by chromatin immunoprecipitation. How hnRNP-K can repress transcription however is unknown and it will be interesting to see how it is specifically recruited at gene promoters.

lncRNAs as Activators of Transcription

Transcription activation by lncRNAs is a much more recent discovery than inhibition. Antisense ncRNAs transcribed intergenically from the *HOXA* cluster were amongst the first transcripts found to induce protein-coding genes. Like HOTAIR, *HOXA*-encoded lncRNAs are tissue-specific, transcribed by RNA pol II, spliced and polyadenylated. However, these transcripts appear to mainly induce the expression of neighboring genes in *cis*. In fact, no lncRNA has yet been shown to induce protein-coding genes in *trans*. Activating lncRNAs have since been reported genome-wide in a number of cell types. As for *HOXA* lncRNAs, they are often tissue-specific and transcribed by RNA pol II. However, some of these transcripts were found to induce distal genes within at least 300 kb away without necessarily affecting the ones adjacent to them. This type of specific long-range regulation suggests that three-dimensional chromatin architecture might be important for their activity. Activation by lncRNAs does not seem to be restricted to protein-coding genes since XCI was recently shown to require this type of mechanism. There will likely be many mechanisms by which lncRNAs can activate transcription (Fig. **3**) although most are currently unknown. Below are examples of lncRNAs thought to play a role in *cis* transcription activation.

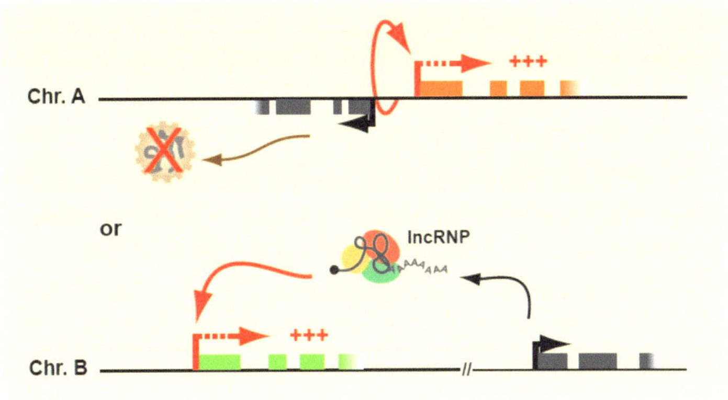

Fig. (3). *lncRNAs as activators of transcription.* (*top*) The process of lncRNA transcription and not the transcript itself could be responsible for *cis* activation of neighboring genes by maintaining or initiating the chromatin structure in an active state. (*bottom*) lncRNAs can be packaged with transcript-specific proteins into lncRNPs to selectively activate gene expression in *cis*. Protein-coding genes are represented by orange and green boxes. lncRNA genes are illustrated in grey. Transcription start sites (TSSs) are indicated by bent arrows. Thick red bent arrows indicate active protein coding genes and thick black arrows active lncRNA genes.

Hox-Encoded ncRNAs, HOTAIRM1, and HOTTIP

An interesting aspect to *HOX* cluster regulation is reflected in the developmental expression pattern of their genes. During development, *HOX* genes are expressed in a spatio-temporal manner that is collinear with their positions along chromosomes [101, 102]. For example, genes located at the 3' end of the clusters are usually expressed more anteriorly and earlier in the embryo than those at the 5' end. Although genetic evidence suggests a role for three-dimensional chromatin organization in collinearity, how this type of *HOX* regulation occurs is unclear [103]. In early 2007, Orlando and colleagues used the well-described NT2/D1 cell differentiation system to investigate the role of intergenic *HOXA* ncRNAs in regulating *cis* gene expression [104]. NT2/D1 are human pluripotent embryonic carcinoma cells that can be terminally differentiated into neural-like cell lineages upon treatment with retinoic acid (RA). RA treatment also recapitulates the 3'-5' induction pattern of the *HOXA1* to *A5* 3' end genes in developing axial systems. The study showed that RA induces antisense *HOXA* ncRNAs in a manner that follows the collinear activation pattern of the protein-coding genes. Importantly, ncRNA induction was also accompanied by local changes in histone modifications and loss of repressive PcG complexes. It was proposed that antisense ncRNA transcription is fundamentally important to open and maintain the active state of *HOX* clusters during RA induction. In this model, the process of transcription itself rather than transcripts *per se* could represent an anti-silencing mechanism. The authors suggested that ncRNA genes might contain regulative epigenetic DNA elements and that transcription through them may counteract repression and maintain RA activation. Although we still do not know which elements could be at these sites, an attractive possibility is that transcription displaces chromatin-looping factors and disrupts repressive three-dimensional conformations. We previously reported the presence of long-range chromatin loops in the transcriptionally silent *HOXA* cluster in THP-1 and NT2/D1 cells [105, 106]. We found that RA treatment abolished these loops in NT2/D1 cells and suggested that the insulator-binding protein CTCF might be an important looping factor [106-109]. It will be interesting to determine the role of CTCF in regulating *HOXA* gene expression.

HOXA ncRNA expression is not restricted to the induction of pluripotent cells with RA. *HOX* cluster transcriptome analysis in human fibroblasts actually identified hundreds of ncRNAs throughout the four clusters [56]. By using cells corresponding to various anatomical sites along developmental axes, this study revealed that ncRNA transcription is often tissue-specific and might demarcate silent and active chromatin domains. A tissue-specific antisense lncRNA transcribed between *HOXA1* and *A2* was more recently identified in the myeloid cell lineage [110, 111]. Because of its myeloid-specific expression, the transcript was termed HOTAIRM1 for "*HOX* Antisense Intergenic RNA Myeloid 1". HOTAIRM1 is 500 nt in length, transcribed by RNA pol II, spliced and polyadenylated. It is the most highly expressed *HOXA* lncRNA product following RA differentiation of NB4 cells into granulocytes and is highly expressed in leukocytes during normal human hematopoiesis. HOTAIRM1 knockdown in NB4 cells prevented RA induction of *HOXA1* and *A4*, and attenuated the induction of the *CD11b* and *CD18* granulocyte maturation genes. These results therefore suggest that transcription activation by HOTAIRM1 may not be restricted to its immediate *HOXA* neighboring genes, and it will be intriguing to determine its role in hematopoiesis and leukemogenesis.

An article recently published by the Chang group also reports a lncRNA at the distal 5' end of the *HOXA* cluster termed HOTTIP for "*HOXA* Transcript at the distal TIP" [112]. This 3,764-nt ncRNA is spliced, polyadenylated, and preferentially expressed at distal anatomical sites where 5' end *HOX* genes are usually expressed. For example, HOTTIP was detected in human foreskin fibroblasts but not in lung. It was also expressed in mouse limbs at E13.5 and in chick limb buds. HOTTIP was shown to positively regulate the expression of 5' *HOXA* genes in human distal fibroblasts and developing chick limbs. Indeed, knockdown of HOTTIP in foreskin fibroblasts resulted in lower 5' *HOXA* gene expression and reduced H3K4me3 at promoters, but did not affect the levels of H3K27me3 along the cluster. Consistent with the role of 5' end *HOXA* genes in limb development, HOTTIP shRNA injection in chicken embryos interfered with the formation of distal bony elements of the wing. These results show that HOTTIP might promote 5' end *HOXA* transcription epigenetically by increasing the levels of activating histone marks. The mixed lineage leukemia proteins MLL1 and MLL2 are primarily responsible for trimethylation of H3K4 at the *HOX* clusters and HOTTIP was found to interact directly with the WDR5 component of the MLL modifying complex. HOTTIP knockdown reduced MLL1 and WDR5 binding at the 5' end *HOX* gene promoters suggesting that the lncRNA is required to recruit the methyltransferase complex to the chromatin. However, ectopic expression of HOTTIP in non-expressing cells did not activate the transcription of 5' end *HOXA* genes. Similarly, HOTTIP overexpression in foreskin fibroblasts did not further enhance *HOXA* gene expression or rescue the effects of depleting endogenous nascent HOTTIP RNA. These results were explained with the suggestion that HOTTIP might only regulate genes when expressed immediately in *cis* of its targets and by hijacking pre-existing higher-order chromosomal structures that group regulated genes in close physical proximity to nascent HOTTIP RNA. Accordingly, 5C analysis of the *HOXA* cluster in fibroblasts indicated that HOTTIP expression correlates with enhanced chromatin interactions at the cluster 5' end between *HOXA13* and *HOXA7*. Frequent chromatin contacts corresponded to higher H3K4me3 levels and low H3K27me3 marks, and HOTTIP knockdown had no effect on three-dimensional chromatin organization. Thus, HOTTIP might represent an example whereby spatial chromatin organization is required to regulate gene expression by lncRNAs in *cis*. We have previously shown that the transcriptionally silent *HOX* clusters are extensively folded in different cell lines [105-107]. Our findings along with the *HOXA* structural data summarized above and the exquisite richness of *HOX* clusters in lncRNAs suggest that the clusters might be regulated in a cell type-specific manner by different mechanisms involving both chromatin architecture and lncRNAs.

Activating lncRNAs (ncRNA-a)

In 2010, Orom and colleagues used GENCODE annotation data to search for uncharacterized human lncRNAs [113, 114]. By focusing on intergenic transcripts classified to have no coding potential by the HAVANA group (Human and Vertebrate Analysis and Annotation; Sanger Institute), they identified 3019 putative long non-coding RNAs. These transcripts displayed chromatin signatures similar to protein-coding genes with high H3K4me3 at their 5' ends and H3K36me3 along their bodies when transcriptionally active. In contrast to mRNAs however, the identified lncRNAs were shorter with an average transcript size of 800 nt. Their genes were also less complex than protein-coding genes with nearly 50% of them featuring a single intron. lncRNA expression analysis on custom arrays indicated that approximately 20% were ubiquitously expressed while over 33% displayed differential expression patterns across cell lines. Furthermore, it was found that the expression of many lncRNAs was regulated during cellular differentiation. Interestingly, functional knockdown analysis revealed an enhancer-like function for a set of lncRNAs. Indeed it was found that depletion of lncRNAs at multiple loci could result in a specific decrease in the expression of protein-coding genes in *cis*. Transcription activation was shown to be RNA-dependent, as evidenced by heterologous transcription assays. Importantly, some of these "ncRNA-a" were found to activate the transcription of critical regulators of development and cellular differentiation. For example, the Snai1 and Snai2 master regulators of hematopoiesis were downregulated following depletion of neighboring lncRNAs. Specifically, ncRNA-a7 RNAi knockdown was accompanied with less Snai1 expression and recapitulated the effects of Snail depletion itself. Interestingly, ncRNA-a7 depletion also led to a decrease in the expression of the *Aurora-kinase A* gene located 6 Mb away from the lncRNA gene. In fact, transcriptional effects were not restricted to immediate neighboring genes for many ncRNA-a, and could extend at least 300 kb away from the transcription units without necessarily affecting all genes in between. Thus, like *HOX*-encoded lncRNAs, these results suggest that spatial chromatin organization might play an important role in the specific activation of target genes.

Jpx Activation of Xist

In addition to activating the transcription of protein-coding genes, lncRNAs can also induce the expression of other lncRNAs. This type of regulation was found to be essential for XCI where Xist expression is repressed by Tsix on the Xa but induced by Jpx on Xi. Jpx is one of the 6 ncRNAs encoded at the Xic that was recently shown to be required for XCI [115]. Specifically, Jpx was found to derive mainly from the Xi and its expression to correlate with Xist transcription. Importantly, deleting Jpx in mouse blocked XCI and was female lethal. This phenotype was rescued by truncating Tsix suggesting that antagonistic lncRNA switches control XCI. Although it can work in *trans*, Jpx was found to have a mild preference for *cis* activation. Interestingly, previous 3C analysis of the Xic revealed higher-order chromatin contacts between the Jpx gene 5' end and the Xist gene [116]. This result suggests that similarly to HOTTIP, three-dimensional chromatin organization might participate in preferentially targeting the Jpx lncRNA to its *cis*-located Xist allele. This study also reinforces an important principle in lncRNA function: that lncRNAs can regulate the expression of other lncRNAs. As such, it seems very likely that this type of regulatory network will play a role in the regulation of protein-coding genes.

ncRNAs as Basic Transcriptional Features

Ascribing a function to long ncRNAs that are capped, spliced, polyadenylated and tissue-specific is intuitively easier to accept than to do the same for low abundance, short, and apparently unprocessed transcripts. In the past, short ncRNAs were often viewed as likely degradation products or as transcriptional "noise", and were routinely discarded from datasets. Several groups have now taken a closer look at these RNAs and show that their non-random distribution is incompatible with aberrant transcription or with the truncated fragments of highly expressed genes. Short ncRNAs tend to be derived specifically from enhancer and promoter regions, and their expression levels, although very low, usually rises with increasing transcription (Fig. **4**). Regardless of whether these transcripts are simple manifestations of transcription activity or have actual regulatory functions, we can certainly learn much about transcription by taking a closer look at them.

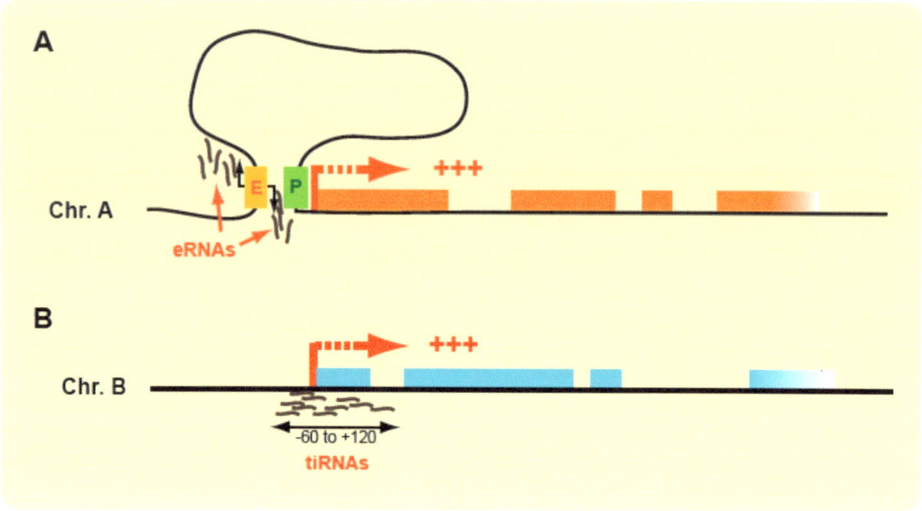

Fig. (4). *ncRNAs as basic transcriptional features.* (**A**) Bi-directional short ncRNAs produced from the center of enhancer sequences can be detected in cortical neurons. These "enhancer RNAs" (eRNAs) are only found near active genes and require RNA pol II. eRNA production would presumably require contact between the enhancer and transcribed gene. (**B**) Short (18 nt) transcription initiation RNAs (tiRNAs) can be detected at active genes. tiRNAs preferentially map around TSSs of highly transcribed genes and sites of RNA pol II binding. Protein-coding genes are represented by orange or blue boxes. Active transcription start sites (TSSs) are indicated thick red bent arrows. Enhancer sequence (E) is shown as a yellow box and promoter (P) as a green rectangle.

Enhancer-Associated RNAs (eRNAs)

Enhancer sequences are scattered throughout mammalian genomes and are often activated in a tissue-specific manner [117]. Active enhancers can be identified by their association with the p300/CBP transcriptional co-activator, by the presence of histone 3 lysine 4 monomethylation marks (H3K4me1), and by their separate position to known transcription start sites (TSSs). Greenberg and colleagues recently used these enhancer traits to identify thousands of regulated enhancers during cortical neuron stimulation [118]. The study reports that most regulated p300/CBP binding sites are at least 1 kb away from known TSSs. Thus, transcription-dependent p300/CBP binding mainly occurs at enhancers rather than promoters in this system. Stimulation did not significantly change the histone modification landscape suggesting that chromatin is likely maintained permissive for induction in these cells. However, many transcription factors (TFs) were found to co-localize at enhancers, thereby raising the possibility that TFs might regulate enhancer function partly by recruiting p300/CBP. Strikingly, roughly 25% of enhancer-bound p300/CBP was found to bind RNA pol II in a transcription-dependent manner. Furthermore, short RNAs (< 2kb) transcribed bi-directionally from the center of enhancer domains could be detected from these sites. These transcripts were termed eRNAs for "enhancer RNAs" and appear to be non-polyadenylated. An interesting feature about eRNA transcription is that their levels correlate with the activity of nearby protein coding genes during neuronal stimulation. As such, enhancers may not strictly be recruitment sites of TFs, p300/CBP and RNA pol II but also sites of transcription. Changes in eRNA expression at actively transcribed enhancers were strongly correlated with regulated mRNA levels at nearby genes, and suggested that eRNA synthesis may only occur when enhancers interact with promoters. This hypothesis was confirmed by the absence of eRNA production from the active *Arc* gene enhancer in cells lacking the *Arc* promoter and most of the gene. Whether the transcriptional process itself or actual eRNA transcripts play a role in enhancer activation by modifying the chromatin state remains unknown. Nonetheless, this study not only clearly demonstrates that enhancers can be transcribed but also indicates that long-range looping rather than transcription-dependent RNA pol II tracking is a likely more widespread mechanism by which distal enhancers regulate target genes.

Transcription initiation RNAs (tiRNAs)

Short ncRNAs mapping nearby gene promoters have been reported in several species over the past five years [119-122]. These studies together led to the hypothesis that genomes might encode multifunctional sequences that sometimes function as regulatory DNA elements and sometimes generate multiple independently regulated transcripts [123]. In 2009, the Mattick group examined small RNA deep-sequencing libraries from human, chicken and *Drosophila* to further characterize TSS-associated transcripts [124]. This study uncovered a new class of short ncRNAs mapping within 60 bp upstream to 120 bp downstream of annotated TSSs that were named tiRNAs for "transcription initiation RNAs". The 5' ends of tiRNAs peaked at 10 to 30 nts downstream from the TSS of their parent genes indicating that they are not simply RNA pol II runoff or truncated 5' capped ends. They were found in all 3 species, and although slightly different in lengths, most were shorter than 22 nt with over a quarter measuring 18 nucleotides. tiRNAs were usually transcribed from the same strand as their corresponding TSS and were preferentially derived from GC-rich promoters. Consistently, promoters with tiRNAs were enriched in RNA pol II and binding of the Sp1 transcription factor. Curiously, tiRNAs were more frequently associated with highly expressed genes but their abundance did not directly correlate with gene expression levels. Furthermore, tiRNA synthesis was independent from the RNAi processing machinery indicating that they are likely derived directly from the RNA pol II activity at promoters. At least in *Drosophila*, tiRNA expression was only weakly associated with stalled transcripts produced at poised promoters. The authors therefore suggested that tiRNAs might represent products of RNA pol II backtracking, a process known to require the elongation and transcription cleavage factor TFIIS [125]. In a more recent paper [125], the same group reported a 40-fold nuclear enrichment of tiRNAs and demonstrated their association with transcription initiation marks [126]. Interestingly, tiRNAs were found to possess a 3'-guanine nucleotide bias in all 3 species, which further supports their non-random origin. tiRNA production is therefore a basic transcriptional feature in metazoans that may function in regulating chromatin modifications to fine-tune transcription.

lncRNAs as Regulators of Spatial Genome Organization

The histone-modifying activity of lncRNPs points to their pivotal role in regulating spatial genome organization. Although lncRNAs have thus far been predominantly associated with repressive complexes that methylate H3K27 or H3K9, it seems safe to predict that many more enzymatic activities will likely associate with these transcripts. Post-translational histone modifications may affect three-dimensional chromatin structure by altering the binding of nucleosomes to the DNA, the contacts between nucleosomes, or by changing the recruitment of non-histone proteins. For example, changes in histone charge introduced by acetylation or phosphorylation might significantly affect *in vivo* nucleosome interactions and chromatin compaction levels. Many protein complexes recognizing specific PTMs have now been identified and usually include enzymatic activities that further modify chromatin. Enzymatic activities include HDACs, HATs, HDMTs, HMTs, ubiquitin ligases, and chromatin remodelers. How histone PTMs can tether histone-modifying complexes has been extensively described elsewhere ([5] and references therein).

At least in chicken and *Drosophila*, there is also evidence that ncRNAs themselves contribute to higher-order genome organization [48]. The Azorin group showed that polyA- RNA associates with chromatin where it represents approximately 2 to 5% of total nucleic acids. When they treated chromatin with RNase A, fragments sedimented less rapidly on sucrose gradients and were more sensitive to micrococcal nuclease. These structural changes were not correlated with any significant alteration in histone composition and were observed at euchromatin and heterochromatin. It was suggested that RNA might stabilize the binding of non-histone proteins to chromatin as was previously shown for HP1 at pericentric heterochromatin [127, 128]. A similar type of mechanism was also recently reported at repressed PcG target genes where short transcripts produced around TSSs can bind PRC2 complexes and might strengthen their association with the genes [129].

Specific ncRNAs may also nucleate the formation of functional nuclear subdomains or be required to maintain their integrity [130, 131]. The ncRNA NEAT1 (Nuclear Enriched Abundant Transcripts 1) is a very good example for this type of activity [132-135]. NEAT1 is a ubiquitously expressed nuclear lncRNA of 3.7 kb that is unspliced but polyadenylated [136]. It is highly expressed and localizes mainly to a type of membraneless nuclear body termed "paraspeckle". The function of paraspeckles is not well defined but thought to represent nuclear retention sites of adenosine-to-inosine (A-I) edited transcripts [137]. NEAT1 knockdown was shown to disrupt paraspeckles and its overexpression to result in a larger number of them [138]. Another lncRNA termed NEAT2 for "Nuclear Enriched Abundant Transcript 2" was found to preferentially localize to splicing speckles, which is another type of nuclear subcompartment enriched in splicing factors [136]. NEAT2 is a 8.7 kb unspliced, polyadenylated transcript with a similar expression pattern to NEAT1 that represent the bulk of polyA+ RNA in speckles. Although NEAT2 does not appear to play a structural role in these bodies as detected by epifluorescence [138, 139], there is evidence suggesting that it may regulate mRNA alternative splicing by altering the ratio of phosphorylated and unphosphorylated SR splicing factor inside the speckles [140].

In addition to their involvement in the function of nuclear subdomains, ncRNPs could also control higher-order chromatin architecture and transcription activity by mediating long-range contacts. There are many ways by which lncRNAs could affect gene expression through long-range interactions. For example, lncRNAs might cluster to form nuclear organization centers that coordinate gene expression. Good candidates for this type of activity are PcG-bound lncRNAs since polycomb complexes are known to form long-range *cis* and *trans* interactions genome-wide [141]. In this case, physical lncRNP networks would facilitate the coordinated down-regulation of multiple genes. Given the increasing complexity of RNA sequences found to interact with PRC2, it will be interesting to see which lncRNAs may be involved in this type of distal regulation. lncRNP-dependent physical networks could also be important for the formation of transcription factories (Fig. **5A**). Alternatively, lncRNA transcription might enhance or inhibit transcription of neighboring genes by displacing proteins from the chromatin and altering physical loops (Fig. **5B**). In this case, it is the transcription process itself rather than the transcripts that would mediate changes in gene expression. The lncRNAs genes themselves may also regulate chromatin architecture. For instance, lncRNA genes might cluster with other genes to promote the formation of transcription factories and gene expression. Additionally, lncRNA genes could compete with other genes (protein-coding or otherwise) for the same enhancers or other type of regulatory DNA element.

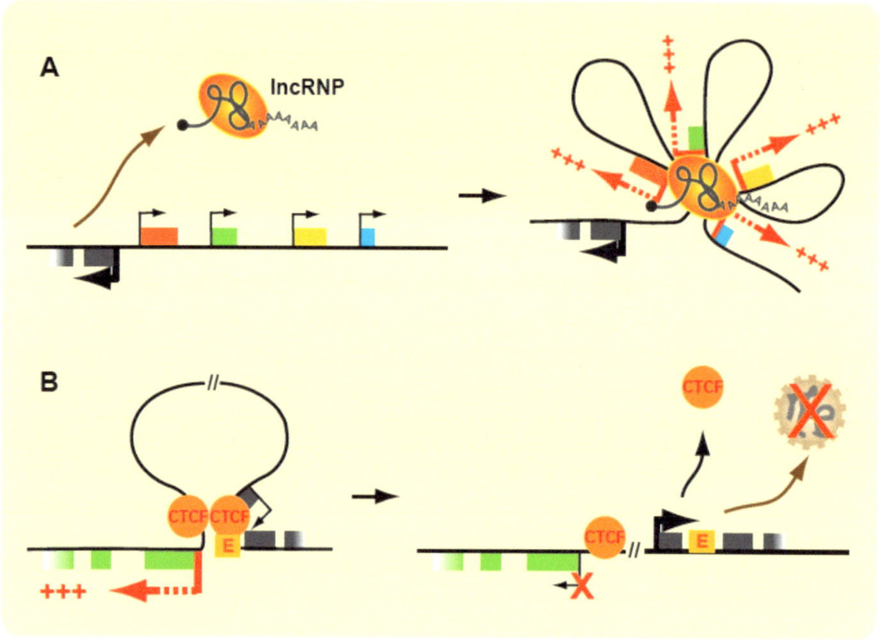

Fig. (5). *lncRNAs as regulators of spatial genome organization.* (**A**) A lncRNP produced in *cis* or in *trans* could drive the formation of activating contacts that bridge enhancers to promoters, or induce the formation of a transcription factory. This example illustrates how a *cis* lncRNA could lead to the coordinated activation of gene clusters. (**B**) Activating long-range looping contacts between enhancers and promoters could be disrupted by transcription of a lncRNA. In this example, lncRNA transcription itself rather than the transcript could be responsible for transcription inhibition of neighboring genes. This example shows one CTCF protein displaced by transcription and disruption of an activating loop. Protein-coding genes are represented by colored boxes. Transcription start sites (TSSs) are indicated by bent arrows, with thick red bent arrows indicating active protein coding genes and thick black arrows active lncRNA genes. TSS of inhibited gene is indicated by a thin bent arrow marked with an "X".

CONCLUSION

The discovery that lncRNAs represent a very large portion of our transcriptome has uncovered a new level of gene expression regulation. Although we know very little about them, their sequence diversity, potential splice variants, and cellular distribution suggests an important role in many aspects of RNA transcription, processing, and metabolism. The tissue-specificity and low conservation of lncRNAs amongst higher eukaryotes also points to their potential contribution in establishing or maintaining cell identity. Because of the overwhelming lncRNA complexity, recent studies have mainly focused on the function of individual transcripts in the nucleus. These studies clearly demonstrate that lncRNAs can regulate genes by various epigenetic mechanisms involving chromatin-modifying complexes. An interesting possibility is that lncRNAs also regulate transcription by altering chromatin architecture. Such spatial transcription regulation mechanisms could play an important role in human health and disease and it will be interesting to see how lncRNA function intersects with spatial chromatin architecture.

CONFLICT OF INTEREST

The author confirms that he has no conflict of interest to declare for this publication.

ACKNOWLEDGEMENTS

We thank members of our laboratory for insightful discussions. We are thankful to Dr. J. Teodoro for critical reading and comments on this manuscript. Research in J.D. lab is funded by grants from the Canadian Institutes of Health Research (CIHR DC0190GP), and the Canadian Cancer Society Research Institute (CCSRI 019252). X.Q.D.W. is supported by a scholarship from the Cole Foundation. J.D. is a CIHR New Investigator and FRSQ Research Scholar (Fonds de la Recherche en Santé du Québec). Funding for open access charge of this article was from the CCSRI 019252 grant to J.D.

DISCLOSURE

This is an updated chapter of the earlier review article [by Wang, X. Q., Crutchley, J. L., & Dostie, J. entitled "Shaping the genome with non-coding RNAs". *Current Genomics*, 2011; 12(5), 307] with additional text and references.

ABBREVIATIONS

3C	=	chromosome conformation capture
eRNAs	=	enhancer RNAs
HMTs	=	histone methyl transferases
Igf2r	=	insulin-like growth factor II receptor
lincRNAs	=	long intergenic non-coding RNAs
lncRNPs	=	long non-coding ribonucleoprotein complexes
miRNAs	=	microRNAs
piRNAs	=	piwi-interacting RNAs
PTMs	=	post-translational modifications
RNAi	=	RNA interference
rRNAs	=	ribosomal RNAs
snoRNAs	=	small nucleolar RNAs
snRNAs	=	small nuclear RNAs
spliRNAs	=	splice-site RNAs
TFs	=	transcription factors
tiRNAs	=	transcription initiation RNAs
tRNAs	=	transfer RNAs
TSSs	=	transcriptional start sites

REFERENCES

[1] Cremer T, Cremer M. Chromosome territories. Cold Spring Harb Perspect Biol. 2010;2(3):a003889.
[2] Pombo A, Branco MR. Functional organisation of the genome during interphase. Curr Opin Genet Dev. 2007;17(5):451-5.
[3] Towbin BD, Meister P, Gasser SM. The nuclear envelope--a scaffold for silencing? Curr Opin Genet Dev. 2009;19(2):180-6.

[4] Li B, Carey M, Workman JL. The role of chromatin during transcription. Cell. 2007;128(4):707-19.
[5] Kouzarides T. Chromatin modifications and their function. Cell. 2007;128(4):693-705.
[6] Calo E, Wysocka J. Modification of enhancer chromatin: what, how, and why? Mol Cell. 2013;49(5):825-37.
[7] Tolhuis B, Palstra RJ, Splinter E, Grosveld F, de Laat W. Looping and Interaction between Hypersensitive Sites in the Active beta-globin Locus. Mol Cell. 2002;10(6):1453-65.
[8] Spilianakis CG, Lalioti MD, Town T, Lee GR, Flavell RA. Interchromosomal associations between alternatively expressed loci. Nature. 2005;435(7042):637-45.
[9] Dean A. In the loop: long range chromatin interactions and gene regulation. Brief Funct Genomics. 2011;10(1):3-10.
[10] Gheldof N, Smith EM, Tabuchi TM, *et al.* Cell-type-specific long-range looping interactions identify distant regulatory elements of the CFTR gene. Nucleic Acids Res. 2010;38(13):4325-36.
[11] Chavanas S, Adoue V, Mechin MC, *et al.* Long-range enhancer associated with chromatin looping allows AP-1 regulation of the peptidylarginine deiminase 3 gene in differentiated keratinocyte. PloS one. 2008;3(10):e3408.
[12] Bugge A, Siersbaek M, Madsen MS, Gondor A, Rougier C, Mandrup S. A novel intronic peroxisome proliferator-activated receptor gamma enhancer in the uncoupling protein (UCP) 3 gene as a regulator of both UCP2 and -3 expression in adipocytes. J Biol Chem. 2010;285(23):17310-7.
[13] Amano T, Sagai T, Tanabe H, Mizushina Y, Nakazawa H, Shiroishi T. Chromosomal dynamics at the Shh locus: limb bud-specific differential regulation of competence and active transcription. Dev Cell. 2009;16(1):47-57.
[14] Lomvardas S, Barnea G, Pisapia DJ, Mendelsohn M, Kirkland J, Axel R. Interchromosomal interactions and olfactory receptor choice. Cell. 2006;126(2):403-13.
[15] Dekker J, Rippe K, Dekker M, Kleckner N. Capturing Chromosome Conformation. Science. 2002;295(5558):1306-11.
[16] Dostie J, Richmond TA, Arnaout RA, *et al.* Chromosome Conformation Capture Carbon Copy (5C): a massively parallel solution for mapping interactions between genomic elements. Genome Research. 2006;16(10):1299-309.
[17] Dostie J, Zhan Y, Dekker J. Chromosome conformation capture carbon copy technology. Curr Protoc Mol Biol. 2007;Chapter 21:Unit 21 14.
[18] Lieberman-Aiden E, van Berkum NL, Williams L, *et al.* Comprehensive Mapping of Long-Range Interactions Reveals Folding Principles of the Human Genome. Science. 2009;326(5950):289-93.
[19] Crutchley JL, Wang XQ, Ferraiuolo MA, Dostie J. Chromatin conformation signatures: ideal human disease biomarkers? Biomark Med. 2010;4(4):611-29.
[20] Dostie J, Bickmore WA. Chromosome organization in the nucleus - charting new territory across the Hi-Cs. Curr Opin Genet Dev. 2012;22(2):125-31.
[21] Vakoc C, Letting DL, Gheldof N, *et al.* Proximity among Distant Regulatory Elements at the beta-Globin Locus Requires GATA-1 and FOG-1. Mol Cell. 2005;17(3):453-62.
[22] Fullwood MJ, Liu MH, Pan YF, *et al.* An oestrogen-receptor-[agr]-bound human chromatin interactome. Nature. 2009;462(7269):58-64.
[23] Yoon H, Boss JM. PU.1 binds to a distal regulatory element that is necessary for B cell-specific expression of CIITA. J Immunol. 2010;184(9):5018-28.
[24] Yochum GS, Sherrick CM, Macpartlin M, Goodman RH. A beta-catenin/TCF-coordinated chromatin loop at MYC integrates 5' and 3' Wnt responsive enhancers. Proc Natl Acad Sci U S A. 2009;107(1):145-50.
[25] Phillips JE, Corces VG. CTCF: master weaver of the genome. Cell. 2009;137(7):1194-211.
[26] Merkenschlager M. Cohesin: a global player in chromosome biology with local ties to gene regulation. Curr Opin Genet Dev. 2010;20(5):555-61.
[27] Millau JF, Gaudreau L. CTCF, cohesin, and histone variants: connecting the genome. Biochemistry and cell biology = Biochimie et biologie cellulaire. 2011;89(5):505-13.
[28] Merkenschlager M, Odom DT. CTCF and cohesin: linking gene regulatory elements with their targets. Cell. 2013;152(6):1285-97.
[29] Rubio ED, Reiss DJ, Welcsh PL, *et al.* CTCF physically links cohesin to chromatin. Proc Natl Acad Sci U S A. 2008;105(24):8309-14.

[30] Hadjur S, Williams LM, Ryan NK, *et al.* Cohesins form chromosomal cis-interactions at the developmentally regulated IFNG locus. Nature. 2009;460(7253):410-3.
[31] Hou C, Dale R, Dean A. Cell type specificity of chromatin organization mediated by CTCF and cohesin. Proc Natl Acad Sci U S A. 2010;107(8):3651-6.
[32] Mishiro T, Ishihara K, Hino S, *et al.* Architectural roles of multiple chromatin insulators at the human apolipoprotein gene cluster. EMBO J. 2009;28(9):1234-45.
[33] Schmidt D, Schwalie PC, Ross-Innes CS, *et al.* A CTCF-independent role for cohesin in tissue-specific transcription. Genome Res. 2010;20(5):578-88.
[34] Kagey MH, Newman JJ, Bilodeau S, *et al.* Mediator and cohesin connect gene expression and chromatin architecture. Nature. 2010;467(7314):430-5.
[35] Nativio R, Wendt KS, Ito Y, *et al.* Cohesin is required for higher-order chromatin conformation at the imprinted IGF2-H19 locus. PLoS Genet. 2009;5(11):e1000739.
[36] Ling JQ, Hou A, Hoffman AR. Long-range DNA interactions are specifically altered by locked nucleic acid-targeting of a CTCF binding site. Biochim Biophys Acta. 2011;1809(1):24-33.
[37] Kurukuti S, Tiwari VK, Tavoosidana G, *et al.* CTCF binding at the H19 imprinting control region mediates maternally inherited higher-order chromatin conformation to restrict enhancer access to Igf2. Proc Natl Acad Sci U S A. 2006;103(28):10684-9.
[38] Li T, Hu JF, Qiu X, *et al.* CTCF regulates allelic expression of Igf2 by orchestrating a promoter-polycomb repressive complex 2 intrachromosomal loop. Mol Cell Biol. 2008;28(20):6473-82.
[39] Kapranov P, Cawley SE, Drenkow J, *et al.* Large-scale transcriptional activity in chromosomes 21 and 22. Science. 2002;296(5569):916-9.
[40] Rinn JL, Euskirchen G, Bertone P, *et al.* The transcriptional activity of human Chromosome 22. Genes Dev. 2003;17(4):529-40.
[41] Bertone P, Stolc V, Royce TE, *et al.* Global identification of human transcribed sequences with genome tiling arrays. Science. 2004;306(5705):2242-6.
[42] Cheng J, Kapranov P, Drenkow J, *et al.* Transcriptional maps of 10 human chromosomes at 5-nucleotide resolution. Science. 2005;308(5725):1149-54.
[43] Carninci P, Kasukawa T, Katayama S, *et al.* The transcriptional landscape of the mammalian genome. Science. 2005;309(5740):1559-63.
[44] Katayama S, Tomaru Y, Kasukawa T, *et al.* Antisense transcription in the mammalian transcriptome. Science. 2005;309(5740):1564-6.
[45] Birney E, Stamatoyannopoulos JA, Dutta A, *et al.* Identification and analysis of functional elements in 1% of the human genome by the ENCODE pilot project. Nature. 2007;447(7146):799-816.
[46] He Y, Vogelstein B, Velculescu VE, Papadopoulos N, Kinzler KW. The antisense transcriptomes of human cells. Science. 2008;322(5909):1855-7.
[47] Clark MB, Mattick JS. Long noncoding RNAs in cell biology. Semin Cell Dev Biol. 2011.
[48] Rodriguez-Campos A, Azorin F. RNA is an integral component of chromatin that contributes to its structural organization. PLoS One. 2007;2(11):e1182.
[49] Chen LL, Carmichael GG. Altered nuclear retention of mRNAs containing inverted repeats in human embryonic stem cells: functional role of a nuclear noncoding RNA. Mol Cell. 2009;35(4):467-78.
[50] Wang K, Yang YW, Liu B, *et al.* A long noncoding RNA maintains active chromatin to coordinate homeotic gene expression. Nature. 2011;In press.
[51] Brown CJ, Ballabio A, Rupert JL, *et al.* A gene from the region of the human X inactivation centre is expressed exclusively from the inactive X chromosome. Nature. 1991;349(6304):38-44.
[52] Brown CJ, Lafreniere RG, Powers VE, *et al.* Localization of the X inactivation centre on the human X chromosome in Xq13. Nature. 1991;349(6304):82-4.
[53] Lee JT. The X as model for RNA's niche in epigenomic regulation. Cold Spring Harb Perspect Biol. 2010;2(9):a003749.
[54] Lessing D, Anguera MC, Lee JT. X chromosome inactivation and epigenetic responses to cellular reprogramming. Annu Rev Genomics Hum Genet. 2013;14:85-110.
[55] Ogawa Y, Lee JT. Xite, X-inactivation intergenic transcription elements that regulate the probability of choice. Mol Cell. 2003;11(3):731-43.
[56] Rinn JL, Kertesz M, Wang JK, *et al.* Functional demarcation of active and silent chromatin domains in human HOX loci by noncoding RNAs. Cell. 2007;129(7):1311-23.

[57] McGinnis W, Garber RL, Wirz J, Kuroiwa A, Gehring WJ. A homologous protein-coding sequence in drosophila homeotic genes and its conservation in other metazoans. Cell. 1984;37(2):403-8.
[58] Dolle P, Izpisua-Belmonte J-C, Falkenstein H, Renucci A, Duboule D. Coordinate expression of the murine Hox-5 complex homoeobox-containing genes during limb pattern formation. Nature. 1989;342(6251):767-72.
[59] Wellik DM, Capecchi MR. Hox10 and Hox11 Genes Are Required to Globally Pattern the Mammalian Skeleton. Science. 2003;301(5631):363-7.
[60] Starkova J, Zamostna B, Mejstrikova E, Krejci R, Drabkin HA, Trka J. HOX gene expression in phenotypic and genotypic subgroups and low HOXA gene expression as an adverse prognostic factor in pediatric ALL. Pediatric Blood & Cancer. 2010;55(6):1072-82.
[61] Wheadon H, Ramsey JM, Dobbin E, *et al.* Differential Hox Expression in Murine Embryonic Stem Cell Models of Normal and Malignant Hematopoiesis. Stem Cells and Development. 2011;0(0):null.
[62] Bernstein BE, Mikkelsen TS, Xie X, *et al.* A Bivalent Chromatin Structure Marks Key Developmental Genes in Embryonic Stem Cells. Cell. 2006;125(2):315-26.
[63] Barber BA, Rastegar M. Epigenetic control of Hox genes during neurogenesis, development, and disease. Annals of Anatomy - Anatomischer Anzeiger. 2010;192(5):261-74.
[64] Tsai M-C, Manor O, Wan Y, *et al.* Long Noncoding RNA as Modular Scaffold of Histone Modification Complexes. Science. 2010;329(5992):689-93.
[65] Gupta RA, Shah N, Wang KC, *et al.* Long non-coding RNA HOTAIR reprograms chromatin state to promote cancer metastasis. Nature. 2010;464(7291):1071-6.
[66] Gil J, Peters G. Regulation of the INK4b-ARF-INK4a tumour suppressor locus: all for one or one for all. Nat Rev Mol Cell Biol. 2006;7(9):667-77.
[67] Pasmant E, Laurendeau I, Héron D, Vidaud M, Vidaud D, Bièche I. Characterization of a Germ-Line Deletion, Including the Entire INK4/ARF Locus, in a Melanoma-Neural System Tumor Family: Identification of ANRIL, an Antisense Noncoding RNA Whose Expression Coclusters with ARF. Cancer Research. 2007;67(8):3963-9.
[68] Folkersen L, Kyriakou T, Goel A, *et al.* Relationship between CAD Risk Genotype in the Chromosome 9p21 Locus and Gene Expression. Identification of Eight New ANRIL Splice Variants. PloS one. 2009;4(11):e7677.
[69] Kotake Y, Nakagawa T, Kitagawa K, *et al.* Long non-coding RNA ANRIL is required for the PRC2 recruitment to and silencing of p15INK4B tumor suppressor gene. Oncogene. 2010.
[70] Yap KL, Li S, Muñoz-Cabello AM, *et al.* Molecular Interplay of the Noncoding RNA ANRIL and Methylated Histone H3 Lysine 27 by Polycomb CBX7 in Transcriptional Silencing of INK4a. Molecular Cell. 2010;38(5):662-74.
[71] Cao R, Wang L, Wang H, *et al.* Role of Histone H3 Lysine 27 Methylation in Polycomb-Group Silencing. Science. 2002;298(5595):1039-43.
[72] Cao R, Tsukada Y-i, Zhang Y. Role of Bmi-1 and Ring1A in H2A Ubiquitylation and Hox Gene Silencing. Molecular Cell. 2005;20(6):845-54.
[73] de Napoles M, Mermoud JE, Wakao R, *et al.* Polycomb Group Proteins Ring1A/B Link Ubiquitylation of Histone H2A to Heritable Gene Silencing and X Inactivation. Developmental cell. 2004;7(5):663-76.
[74] Stock JK, Giadrossi S, Casanova M, *et al.* Ring1-mediated ubiquitination of H2A restrains poised RNA polymerase II at bivalent genes in mouse ES cells. Nat Cell Biol. 2007;9(12):1428-35.
[75] Blackledge NP, Farcas AM, Kondo T, *et al.* Variant PRC1 complex-dependent H2A ubiquitylation drives PRC2 recruitment and polycomb domain formation. Cell. 2014;157(6):1445-59.
[76] Yu W, Gius D, Onyango P, *et al.* Epigenetic silencing of tumour suppressor gene p15 by its antisense RNA. Nature. 2008;451(7175):202-6.
[77] Monk M. Genomic imprinting. Genes & Development. 1988;2(8):921-5.
[78] Wutz A, Smrzka OW, Schweifer N, Schellander K, Wagner EF, Barlow DP. Imprinted expression of the Igf2r gene depends on an intronic CpG island. Nature. 1997;389(6652):745-9.
[79] Sleutels F, Zwart R, Barlow DP. The non-coding Air RNA is required for silencing autosomal imprinted genes. Nature. 2002;415(6873):810-3.
[80] Pandey RR, Mondal T, Mohammad F, *et al.* Kcnq1ot1 Antisense Noncoding RNA Mediates Lineage-Specific Transcriptional Silencing through Chromatin-Level Regulation. Molecular Cell. 2008;32(2):232-46.

[81] Lyle R, Watanabe D, Vruchte Dt, *et al.* The imprinted antisense RNA at the Igf2r locus overlaps but does not imprint Mas1. Nat Genet. 2000;25(1):19-21.
[82] Zwart R, Sleutels F, Wutz A, Schinkel AH, Barlow DP. Bidirectional action of the Igf2r imprint control element on upstream and downstream imprinted genes. Genes & Development. 2001;15(18):2361-6.
[83] Seidl CIM, Stricker SH, Barlow DP. The imprinted Air ncRNA is an atypical RNAPII transcript that evades splicing and escapes nuclear export. EMBO J. 2006;25(15):3565-75.
[84] Nagano T, Mitchell JA, Sanz LA, *et al.* The Air Noncoding RNA Epigenetically Silences Transcription by Targeting G9a to Chromatin. Science. 2008;322(5908):1717-20.
[85] Tachibana M, Sugimoto K, Fukushima T, Shinkai Y. SET Domain-containing Protein, G9a, Is a Novel Lysine-preferring Mammalian Histone Methyltransferase with Hyperactivity and Specific Selectivity to Lysines 9 and 27 of Histone H3. Journal of Biological Chemistry. 2001;276(27):25309-17.
[86] Tachibana M, Ueda J, Fukuda M, *et al.* Histone methyltransferases G9a and GLP form heteromeric complexes and are both crucial for methylation of euchromatin at H3-K9. Genes & Development. 2005;19(7):815-26.
[87] Rice JC, Briggs SD, Ueberheide B, *et al.* Histone Methyltransferases Direct Different Degrees of Methylation to Define Distinct Chromatin Domains. Molecular Cell. 2003;12(6):1591-8.
[88] Mitsuya K, Meguro M, Lee MP, *et al.* LIT1, an Imprinted Antisense RNA in the Human KvLQT1 Locus Identified by Screening for Differentially Expressed Transcripts Using Monochromosomal Hybrids. Human Molecular Genetics. 1999;8(7):1209-17.
[89] Bliek J, Maas SM, Ruijter JM, *et al.* Increased tumour risk for BWS patients correlates with aberrant H19 and not KCNQ1OT1 methylation: occurrence of KCNQ1OT1 hypomethylation in familial cases of BWS. Human Molecular Genetics. 2001;10(5):467-76.
[90] Ohya S, Sergeant GP, Greenwood IA, Horowitz B. Molecular Variants of KCNQ Channels Expressed in Murine Portal Vein Myocytes: A Role in Delayed Rectifier Current. Circ Res. 2003;92(9):1016-23.
[91] Thakur N, Tiwari VK, Thomassin H, *et al.* An Antisense RNA Regulates the Bidirectional Silencing Property of the Kcnq1 Imprinting Control Region. Mol Cell Biol. 2004;24(18):7855-62.
[92] Kanduri C, Thakur N, Pandey RR. The length of the transcript encoded from the Kcnq1ot1 antisense promoter determines the degree of silencing. EMBO J. 2006;25(10):2096-106.
[93] Mohammad F, Pandey RR, Nagano T, *et al.* Kcnq1ot1/Lit1 Noncoding RNA Mediates Transcriptional Silencing by Targeting to the Perinucleolar Region. Mol Cell Biol. 2008;28(11):3713-28.
[94] Lane DP. p53, guardian of the genome. Nature. 1992;358(6381):15-6.
[95] Vousden KH, Lu X. Live or let die: the cell's response to p53. Nat Rev Cancer. 2002;2(8):594-604.
[96] Goh AM, Coffill CR, Lane DP. The role of mutant p53 in human cancer. The Journal of Pathology. 2011;223(2):116-26.
[97] Gurpinar E, Vousden KH. Hitting cancers' weak spots: vulnerabilities imposed by p53 mutation. Trends Cell Biol. 2015.
[98] Huarte M, Guttman M, Feldser D, *et al.* A Large Intergenic Noncoding RNA Induced by p53 Mediates Global Gene Repression in the p53 Response. Cell. 2010;142(3):409-19.
[99] Mikkelsen TS, Ku M, Jaffe DB, *et al.* Genome-wide maps of chromatin state in pluripotent and lineage-committed cells. Nature. 2007;448(7153):553-60.
[100] Guttman M, Amit I, Garber M, *et al.* Chromatin signature reveals over a thousand highly conserved large non-coding RNAs in mammals. Nature. 2009;458(7235):223-7.
[101] Krumlauf R. Hox genes in vertebrate development. Cell. 1994;78(2):191-201.
[102] Duboule D, Morata G. Colinearity and functional hierarchy among genes of the homeotic complexes. Trends Genet. 1994;10(10):358-64.
[103] Duboule D, Deschamps J. Colinearity loops out. Developmental cell. 2004;6(6):738-40.
[104] Sessa L, Breiling A, Lavorgna G, Silvestri L, Casari G, Orlando V. Noncoding RNA synthesis and loss of Polycomb group repression accompanies the colinear activation of the human HOXA cluster. Rna. 2007;13(2):223-39.
[105] Fraser J, Rousseau M, Shenker S, *et al.* Chromatin conformation signatures of cellular differentiation. Genome Biol. 2009;10(4):R37.

[106] Ferraiuolo MA, Rousseau M, Miyamoto C, *et al.* The three-dimensional architecture of Hox cluster silencing. Nucleic Acids Res. 2010;38(21):7472-84.
[107] Rousseau M, Crutchley JL, Miura H, Suderman M, Blanchette M, Dostie J. Hox in motion: tracking HoxA cluster conformation during differentiation. Nucleic Acids Res. 2014;42(3):1524-40.
[108] Xu M, Zhao GN, Lv X, *et al.* CTCF controls HOXA cluster silencing and mediates PRC2-repressive higher-order chromatin structure in NT2/D1 cells. Mol Cell Biol. 2014;34(20):3867-79.
[109] Narendra V, Rocha PP, An D, Raviram R, Skok JA, Mazzoni EO, et al. Transcription. CTCF establishes discrete functional chromatin domains at the Hox clusters during differentiation. Science. 2015;347(6225):1017-21.
[110] Zhang X, Lian Z, Padden C, *et al.* A myelopoiesis-associated regulatory intergenic noncoding RNA transcript within the human HOXA cluster. Blood. 2009;113(11):2526-34.
[111] Zhang X, Weissman SM, Newburger PE. Long intergenic non-coding RNA HOTAIRM1 regulates cell cycle progression during myeloid maturation in NB4 human promyelocytic leukemia cells. RNA biology. 2014;11(6):777-87.
[112] Wang KC, Yang YW, Liu B, *et al.* A long noncoding RNA maintains active chromatin to coordinate homeotic gene expression. Nature. 2011;472(7341):120-4.
[113] Harrow J, Denoeud F, Frankish A, *et al.* GENCODE: producing a reference annotation for ENCODE. Genome Biol. 2006;7 Suppl 1:S4 1-9.
[114] Orom UA, Derrien T, Beringer M, *et al.* Long noncoding RNAs with enhancer-like function in human cells. Cell. 2010;143(1):46-58.
[115] Tian D, Sun S, Lee JT. The long noncoding RNA, Jpx, is a molecular switch for X chromosome inactivation. Cell. 2010;143(3):390-403.
[116] Tsai CL, Rowntree RK, Cohen DE, Lee JT. Higher order chromatin structure at the X-inactivation center via looping DNA. Dev Biol. 2008;319(2):416-25.
[117] Shiekhattar R. Opening the Chromatin by eRNAs. Mol Cell. 2013;51(5):557-8.
[118] Kim TK, Hemberg M, Gray JM, *et al.* Widespread transcription at neuronal activity-regulated enhancers. Nature. 2010;465(7295):182-7.
[119] Davis CA, Ares M, Jr. Accumulation of unstable promoter-associated transcripts upon loss of the nuclear exosome subunit Rrp6p in Saccharomyces cerevisiae. Proc Natl Acad Sci U S A. 2006;103(9):3262-7.
[120] Chekanova JA, Gregory BD, Reverdatto SV, *et al.* Genome-wide high-resolution mapping of exosome substrates reveals hidden features in the Arabidopsis transcriptome. Cell. 2007;131(7):1340-53.
[121] Guenther MG, Levine SS, Boyer LA, Jaenisch R, Young RA. A chromatin landmark and transcription initiation at most promoters in human cells. Cell. 2007;130(1):77-88.
[122] Morris KV, Mattick JS. The rise of regulatory RNA. Nat Rev Genet. 2014;15(6):423-37.
[123] Kapranov P, Willingham AT, Gingeras TR. Genome-wide transcription and the implications for genomic organization. Nat Rev Genet. 2007;8(6):413-23.
[124] Taft RJ, Glazov EA, Cloonan N, *et al.* Tiny RNAs associated with transcription start sites in animals. Nat Genet. 2009;41(5):572-8.
[125] Pal M, McKean D, Luse DS. Promoter clearance by RNA polymerase II is an extended, multistep process strongly affected by sequence. Mol Cell Biol. 2001;21(17):5815-25.
[126] Taft RJ, Simons C, Nahkuri S, *et al.* Nuclear-localized tiny RNAs are associated with transcription initiation and splice sites in metazoans. Nat Struct Mol Biol. 2010;17(8):1030-4.
[127] Maison C, Bailly D, Peters AH, *et al.* Higher-order structure in pericentric heterochromatin involves a distinct pattern of histone modification and an RNA component. Nat Genet. 2002;30(3):329-34.
[128] Muchardt C, Guilleme M, Seeler JS, Trouche D, Dejean A, Yaniv M. Coordinated methyl and RNA binding is required for heterochromatin localization of mammalian HP1alpha. EMBO Rep. 2002;3(10):975-81.
[129] Kanhere A, Viiri K, Araujo CC, *et al.* Short RNAs are transcribed from repressed polycomb target genes and interact with polycomb repressive complex-2. Mol Cell. 2010;38(5):675-88.
[130] Ip JY, Nakagawa S. Long non-coding RNAs in nuclear bodies. Dev Growth Differ. 2012;54(1):44-54.
[131] Nakagawa S, Hirose T. Paraspeckle nuclear bodies--useful uselessness? Cell Mol Life Sci. 2012;69(18):3027-36.

[132] Chen LL, Carmichael GG. Decoding the function of nuclear long non-coding RNAs. Curr Opin Cell Biol. 2010;22(3):357-64.
[133] Naganuma T, Nakagawa S, Tanigawa A, Sasaki YF, Goshima N, Hirose T. Alternative 3'-end processing of long noncoding RNA initiates construction of nuclear paraspeckles. EMBO J. 2012;31(20):4020-34.
[134] Hirose T, Virnicchi G, Tanigawa A, *et al.* NEAT1 long noncoding RNA regulates transcription via protein sequestration within subnuclear bodies. Mol Biol Cell. 2014;25(1):169-83.
[135] West JA, Davis CP, Sunwoo H, *et al.* The long noncoding RNAs NEAT1 and MALAT1 bind active chromatin sites. Mol Cell. 2014;55(5):791-802.
[136] Hutchinson JN, Ensminger AW, Clemson CM, Lynch CR, Lawrence JB, Chess A. A screen for nuclear transcripts identifies two linked noncoding RNAs associated with SC35 splicing domains. BMC Genomics. 2007;8:39.
[137] Fox AH, Lam YW, Leung AK, *et al.* Paraspeckles: a novel nuclear domain. Curr Biol. 2002;12(1):13-25.
[138] Clemson CM, Hutchinson JN, Sara SA, *et al.* An architectural role for a nuclear noncoding RNA: NEAT1 RNA is essential for the structure of paraspeckles. Mol Cell. 2009;33(6):717-26.
[139] Nakagawa S, Ip JY, Shioi G, *et al.* Malat1 is not an essential component of nuclear speckles in mice. Rna. 2012;18(8):1487-99.
[140] Tripathi V, Ellis JD, Shen Z, *et al.* The nuclear-retained noncoding RNA MALAT1 regulates alternative splicing by modulating SR splicing factor phosphorylation. Mol Cell. 2010;39(6):925-38.
[141] Tiwari VK, Cope L, McGarvey KM, Ohm JE, Baylin SB. A novel 6C assay uncovers Polycomb-mediated higher order chromatin conformations. Genome Res. 2008;18(7):1171-9.

CHAPTER 5

Directed Evolution of Fungal Laccases: An Update

Diana M. Mate[1] and Miguel Alcalde[2,*]

[1]*DWI-Leibniz Institute for Interactive Materials, Forckenbeckstraße 50, 52056 Aachen, Germany and* [2]*Department of Biocatalysis, Institute of Catalysis, CSIC, Cantoblanco, 28049 Madrid, Spain*

Abstract: Fungal laccases are generalists biocatalysts with potential applications that range from bioremediation to novel green processes. Fuelled by molecular oxygen, these enzymes can act on dozens of molecules of different chemical nature, and with the help of redox mediators, their spectrum of oxidizable substrates is further pushed towards xenobiotic compounds (pesticides, industrial dyes, PAHs), biopolymers (lignin, starch, cellulose) and other complex molecules. In recent years, extraordinary efforts have been made to engineer fungal laccases by directed evolution and semi-rational approaches to improve their functional expression or stability. All these studies have taken advantage of *Saccharomyces cerevisiae* as a heterologous host, not only to secrete the enzyme but also, to emulate the introduction of genetic diversity through *in vivo* DNA recombination. Here, we discuss all these endeavours to convert fungal laccases into valuable biomolecular platforms on which new functions can be tailored by directed evolution.

Keywords: Directed evolution, functional expression, fungal laccases, high-throughput screening, *in vivo* DNA recombination, *Saccharomyces cerevisiae*, stability.

1. LACCASES: GENERAL FEATURES

Laccases (EC 1.10.3.2) are typically extracellular monomeric glycoproteins that belong to the blue multicopper oxidase family (together with ascorbate oxidase, bilirubin oxidase and ceruloplasmin, among others) [1-3]. Laccases are considered to be ideal "*green catalysts*" since they are capable of oxidizing a wide variety of compounds in a straightforward manner, using O_2 from the air and releasing H_2O as the only by-product [4-6]. These enzymes harbour one paramagnetic T1 copper (producing the characteristic blue-greenish colour in the oxidized resting state) where the oxidation of the reducing substrate takes place, and a trinuclear T2/T3 copper cluster in which oxygen is reduced to water. The reaction mechanism acts like a battery, storing electrons from the monovalent oxidation of the four

*Corresponding author Miguel Alcalde:** Department of Biocatalysis, Institute of Catalysis, CSIC, Cantoblanco, 28049 Madrid, Spain; Tel: +34 915854806; Fax: +34 915854760; E-mail: malcalde@icp.csic.es.

reducing substrates required to reduce one molecule of oxygen to two molecules of water [3, 7-9]. Individually, laccases catalyze the oxidation of a wide range of aromatic compounds: ortho- and para-diphenols, methoxy-substituted phenols, aromatic amines, benzenothiols, hydroxyindols or syringaldazine. In addition, inorganic/organic metal compounds can also serve as substrates of laccase (*e.g.* Mn^{2+} can be oxidized to Mn^{3+}) [3]. Moreover, the range of reducing substrates can be further expanded to non-phenolic compounds (that are otherwise difficult to oxidize) by including redox mediators from synthetic or natural sources in the so-called laccase-mediator systems. Upon oxidation by laccases, redox mediators oxidize other compounds by non-enzymatic mechanisms, thereby allowing substrates with higher redox potential than laccases to be transformed (*e.g.* non-phenolic lignin dimers or polycyclic aromatic hydrocarbons, PAHs). Acting as diffusible electron carriers, these mediators can oxidize large polymers such as lignin, cellulose or starch, circumventing the inherent problems of enzyme hindrance normally associated to these substrates [10].

Laccases can be found in plants, fungi, bacteria, lichens, oysters and a few insects [11]. Plant and bacterial laccases belong to the group of low-redox potential laccases (with redox potential at the T1 site [E^{o}_{T1}] around +400 mV *vs.* normal hydrogen electrode (NHE)), and in particular the latter are of especial interest since they generally display high thermal stability, a quality really appreciated for industrial settings. However, their practical application is still reduced due to the limited range of substrates that they are able to oxidize. On the other hand, bacterial laccases are useful models for protein engineers, who can take advantage of the ease of manipulation of the bacterial host to carry out rational design or directed evolution. Indeed, several studies have been published in recent years dealing with laboratory evolution of bacterial laccases. The CotA from *Bacillus subtilis* was engineered by directed evolution and structure-based methods with the aim of improving substrate specificity [12-14]. Also, the CotA laccase from *Bacillus licheniformis* was subjected to random and site-directed mutagenesis and the mutants were screened for enhancing functional expression in *E. coli* [15]. Lastly, the metagenome-derived alkaline laccase Lac591 was evolved to obtain better performing laccases for decolorization of textile dyes [16].

Fungal laccases have been studied comprehensively because of their strong catalytic capacities and in particular, there have been many studies into directed evolution of fungal laccases in the last few years. This interest initially stemmed from the higher redox potential of the enzyme that permits a broader array of substrates to be oxidised. Indeed, several laccases from white rot fungi (involved in lignin combustion) possess an E^{o}_{T1} close to +800 mV [2, 11]. Thus, fungal laccases, and especially high-redox potential laccases (HRPLs), may serve: i) to produce 2nd generation biofuels (*i.e.* bioethanol) from lignocellulosic material (*i.e.* laccase and/or laccase-mediator system can oxidize the lignin content of agricultural wastes), or in the manufacture of new products from starch, cellulose and lignin with high added value; ii) in the food industry to process drinks and in bakery products; iii) in the paper industry for pulp-kraft biobleaching, the manufacture of mechanical pulps at low energy cost and to treat effluents; iv) in the textile industry for the remediation of dyes in effluents and textile bleaching (*i.e.* jeans); v) in nanobiotechnology, since laccases belong to the exclusive group of enzymes capable of accepting electrons directly from a cathodic compartment which enables them to be used in the engineering of biosensors (for phenols, oxygen, azides, morphine, codein, catecholamines or flavonoids) with clinical and environmental applications, or in biofuel cells; vi) in the bioremediation through the oxidation of polycyclic aromatic hydrocarbons (PAHs), dioxines, halogenate compounds, phenolic compounds, bencenic derivatives, nitroaromatic compounds and dyes; and vii) in the organic synthesis of complex polymers, drugs, antibiotics and cosmetics [4, 17-20].

There have been many attempts to use rational approaches to engineer fungal laccases over the last couple of decades. In pioneering research, several residues in the neighborhood of the catalytic copper ions were subjected to site directed mutagenesis in order to determine what parameters define the catalytic activity and the redox potentials of these enzymes [21-23]. One consequence of these comprehensive structure-function studies was the generation of a collection of mutants with structural perturbations at the T1 copper site. In addition, directed evolution of fungal laccases (especially HRPLs) has received a recent boost by overcoming the obstacles associated with their functional expression in hosts suitable for *in vitro* evolution experiments [24-26].

2. DIRECTED EVOLUTION OF FUNGAL LACCASES

For most of us, directed evolution represents an elegant shortcut to tailor enzymes with improved features. By mimicking the Darwinist algorithm of natural selection through iterative steps of random mutagenesis and/or DNA recombination, the temporal scale of evolution can be collapsed from millions of years into months rather than weeks of bench work [27-29]. In general, it is important to bear in mind three essential aspects when performing laboratory evolution experiments:

i) It is necessary to have a reliable and sensitive screening assay to identify the small improvements obtained in each round of evolution, generally from 2 to 10-fold improvements per evolutive cycle. In the last years, colorimetric high-throughput screening assays specifically designed for laccase evolution have emerged. All these assays are based on known natural or surrogate substrates of different chemical nature and complexity (from phenols to recalcitrant compounds: ABTS, 2,6-dimethoxyphenol, syringaldazine, acetosyringone, sinapic acid, violuric acid, iodide, anthracene or Poly R478 have been used, among others, to screen mutant laccase libraries) [24, 30-34]. Depending on the approach, the screening assays can be combined in an attempt to enhance several features at once (*e.g.* activity and stability [25] or to avoid the laccase becoming dependent of one specific substrate during evolution [24, 26]).

ii) Diversity should be generated by random mutagenesis and *in vivo* or *in vitro* DNA recombination protocols [35]. Other approaches such as circular permutation, combinatorial saturation mutagenesis and the combination of rational design with directed evolution are also frequently included in the evolutionary strategy, generally yielding good results [24, 36, 37].

iii) It must be possible to functionally express the genetic products with the desired traits. Although *Escherichia coli* is the preferred host organism for directed evolution experiments, the broad differences between the eukaryotic expression system of fungal laccases and that of bacteria (codon usage, missing chaperones, post-translational modifications such as glycosylation or the formation of disulphide bridges, and copper uptake) are shortcomings that are not easily overcome. In fact, all attempts to functionally express fungal laccases in bacteria have resulted in misfolding and the formation of inclusion bodies. Alternatively, the secretory machinery of *Saccharomyces cerevisiae* permits post-translational modifications, and it is also an excellent host to carry out laboratory evolution experiments [38].

2.1. *Saccharomyces cerevisiae* for Directed Laccase Evolution: A Biomolecular Tool Box for the Generation of Diversity

There are four main reasons to use *S. cerevisiae* for the laboratory evolution of fungal laccases:

i) High transformation efficiencies: 10^7-10^8 colony forming units (CFU) per μg of DNA, which is comparable to that of *E. coli* (10^8-10^{10} CFU per μg of DNA) [39].

ii) Episomal vectors are available that facilitates the recovery of interesting mutants without integration into the genomic DNA as generally happens in yeasts such as *Pichia pastoris*. This hurdle was overcome by using the episomal plasmid pBGP1, which contains the constitutive glyceraldehyde 3-phosphate dehydrogenase (GAP) gene promoter, the α-factor prepro-leader to direct protein secretion and the *P. pastoris* autonomous replication sequence PARS1 to maintain the vector as an episomal element [40]. The pBGP1 vector was successfully used to screen xylanase and lipase mutant libraries [40, 41]. Moreover, engineering strategies for the evolution of hydroxynitrile lyases in *P. pastoris* were designed involving the integration of linear expression cassettes to construct and express mutant libraries [42]. Still, the lower transformation efficiencies along with the lack of versatile vectors when comparing with *S. cerevisiae* are main reasons why *P. pastoris* has been hardly used in directed evolution campaigns.

iii) The glycosylation and secretion of laccases avoids the tedious and cumbersome lysis steps required for bacteria. In addition, working with supernatants makes the validation of the screening assay easier since there is much less interference than with complex lysate mixtures.

iv) And last but not least, a high level of homologous DNA recombination can be achieved, which enables *in vivo* shuffled mutant libraries to be produced or the development of new tools to generate diversity [35, 43, 44].

One of the first directed evolution experiments using *in vivo* DNA shuffling was carried out to engineer a low redox potential peroxidase from *Coprinopsis cinerea* with oxidative stability [45]. This pioneering work opened an array of possibilities that led to many researchers beginning to develop new strategies of DNA recombination [35, 46-48]. Taking advantage of yeast physiology, our group has designed several *in vivo* DNA recombination methods (IVOE, IvAM) with the aim of generating suitable crossover events or varying the mutational bias in the framework of *in vitro* laccase evolution, Fig. (**1**). Sequence splicing by IVOE (*In Vivo* Overlap Extension) is a robust and reliable method through which combinatorial saturation mutagenesis, deletion and/or insertion mutagenesis, site-directed recombination or site-directed mutagenesis can be accomplished straightforwardly [35, 44]. The method is based on the engineering of mutagenized primers that contain suitable overhangs, with homologous regions that anneal to each other to generate an autonomously replicating vector containing the mutant gene/s, Fig. (**1A**). This strategy mimics the classical SOE (Sequence Overlap Extension) but it removes several steps, including the *in vitro* cloning. IVOE has been employed to construct mutant libraries for directed evolution of ascomycete and basidiomycete laccases, as well as to carry out semi-rational studies (*i.e.* combinatorial saturation mutagenesis coupled to high-throughput screenings [24, 36]). IvAM (*In vivo* Assembly of Mutant libraries) is another approach that has been successfully used to engineer fungal laccases for improved organic co-solvent tolerance, Fig. (**1B**) [43, 49]. Generally, error-prone PCR (*ep*PCR) methods are unsatisfactory because they are associated with a limited and predicted mutational spectrum derived from the intrinsic bias of each DNA-polymerase. To overcome this problem, the mutation biases of different polymerases can be combined by alternating between them in successive rounds of evolution. IvAM has allowed us to explore the laccase sequence space through the *in vivo* DNA shuffling of several mutant libraries with different mutational spectra in a single round of evolution [24, 43, 49, 50]. More recently, our group has developed a domain mutagenesis method called MORPHING (Mutagenic Organized Recombination Process by Homologous *IN vivo* Grouping) that pursues the introduction and *in vivo* recombination of point mutations in defined DNA fragments, Fig. (**1C**) [51]. MORPHING consists in subjecting specific segments of the target gene to *ep*PCR while the remaining fragments are amplified with a high-fidelity DNA polymerase. Next, all fragments -amplified with homologous overhangs at 5′ and 3′ ends are co-transformed along with the linearized vector in *S. cerevisiae* for *in vivo* recombination and cloning in a single step. This strategy has been successfully used to enhance oxidative stability of a versatile peroxidase [51, 52] and the secretion of an unspecific peroxygenase in yeast [51, 53], as well as to discover consensus-ancestor mutations in aryl alcohol oxidases [54]. MORPHING could be also useful to screen laccase libraries for higher E^{o}_{T1} by random mutagenesis on DNA segments corresponding to protein sequences close to the T1 Cu predicted by quantum mechanics molecular mechanics simulations (unpublished material).

It is also possible to bring together strategies for *in vitro* and *in vivo* DNA recombination to evolve enzymes in the laboratory. For example, CLERY (Combinatorial Libraries Enhanced by Recombination in Yeast) combines *in vitro* and *in vivo* DNA shuffling, Fig. (**1D**) [55]. This method allowed us to generate and screen small mutant libraries of chimeric HRPLs by using as parent types the final evolved variants of the ligninolytic basidiomycetes PM1 and *Pycnoporus cinnabarinus* [56]. In a similar approach, we have combined mutagenic StEP with *in vivo* DNA shuffling to evolve ligninolytic peroxidases, Fig. (**1E**) [57]. There is also an interesting report on how to engineer chimeric fungal laccases from *Trametes* C30 by *in vivo* DNA shuffling [48], where a low redox potential laccase gene was used in all the chimeric libraries to guarantee functional expression.

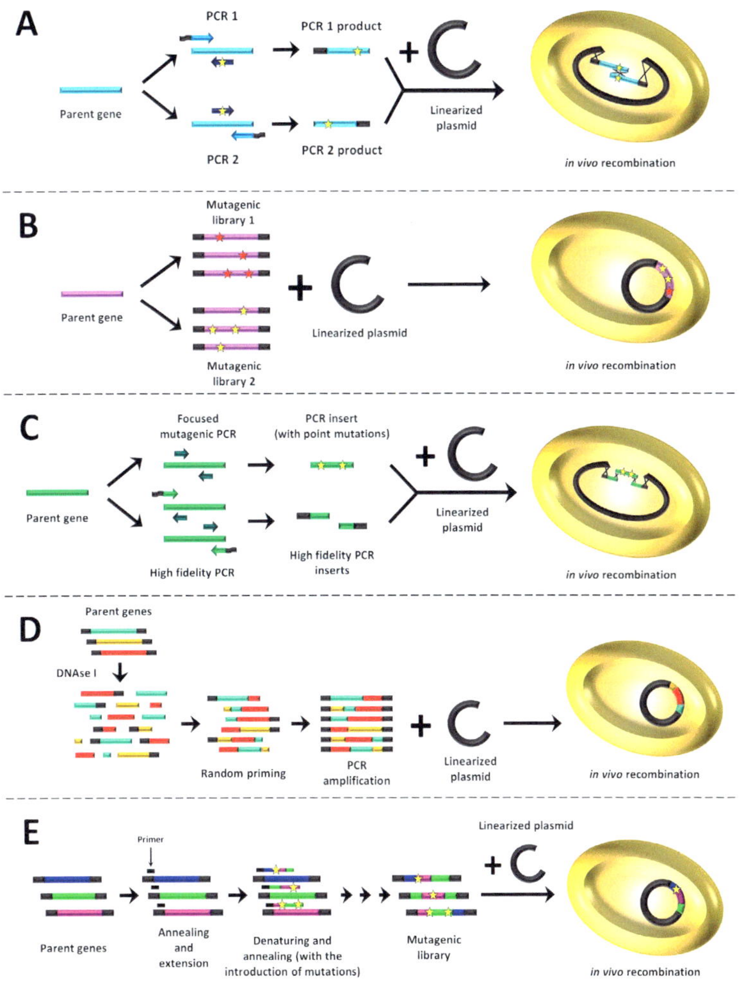

Fig. (1). Different methods used to generate mutant libraries in *S. cerevisiae*: **A**) IVOE; **B**) IvAM; **C**) MORPHING; **D**) CLERY; **E**) StEP+*in vivo* DNA shuffling. Nucleotide substitutions are depicted as stars.

2.2. Directed Evolution of Ascomycete Laccases

The first laccase gene subjected to directed evolution was the *Myceliophthora thermophila* laccase (MtL), a low-medium redox potential ascomycete laccase that is very thermostable (with T_{50} values ~75.6ºC) [46]. In this work, 10 rounds of laboratory evolution were carried out to achieve the strongest functional expression of a laccase in *S. cerevisiae* yet reported (up to 18 mg/L, Fig. **2**). The basic tools for the generation of diversity included *ep*PCR, StEP and *in vivo* DNA shuffling. The latter was modified in such a manner that *ep*PCR products were recombined *in vivo* to introduce new mutations in conjunction with recombination. Furthermore, backcrossing recombination was employed to eliminate neutral mutations. In the final rounds of evolution, PCR and *in vivo* gap repair were used to recombine neighbouring mutations in a site-directed fashion, once again taking advantage of the eukaryotic apparatus (referred to as *in vivo* assembly recombination) which proved extremely useful to eliminate some deleterious mutations. The sequence targeted for directed evolution included the native prepro-leader, as well as the C-terminal tail of the gene that encode for parts of the protein that are cleaved during maturation. The ultimate evolutionary product obtained after screening over 20,000 clones, the T2 mutant, harboured 14 mutations, Fig. (**3A**). The single most beneficial mutation was found in the C-terminal tail, introducing a cleavage site for the *Kex2* protease of the Golgi compartment of *S. cerevisiae*. This mutation probably adjusted the sequence to the different protease specificity of the heterologous host. In a later study, we employed the T2 evolved mutant to study the role of the C-terminal plug of ascomycete laccases. Using combinatorial saturation mutagenesis through IVOE, a direct relationship between the C-terminal plug and a conserved tripeptide in the vicinity of the reducing substrate binding site was determined [36].

Many applications of fungal laccases (bioremediation, lignocellulose processing, organic synthesis, *etc.*) require high concentrations of organic co-solvents in which laccases may unfold and lose their activity. In another study, we used the evolved MtL (T2 mutant) as our point of departure to confer organic co-solvent tolerance [49]. In this work, five rounds of laboratory evolution were carried out to explore 13,000 clones, using IVOE, IvAM and *ep*PCR, in combination with *in vivo* DNA shuffling for library creation. The laboratory evolution strategy was carefully planned according to the following rules:

1. Screening was performed (based on the oxidation of ABTS) in the presence of two co-solvents of different chemical nature and polarity in order to identify variants active in both co-solvents. The ultimate goal was to provide promiscuity in other water-solvent mixtures.

2. To improve the activity and stability in organic co-solvents, only those mutant *hits* that retained their stability and that were more active were considered as candidates for further cycles of evolution.

3. The extreme selective pressure, increasing the concentrations of organic co-solvents from 20% in the first cycles up to 60% in the last round, further drove the evolution.

4. The spectro-electrochemistry of the variants was studied in depth to assess the influence of the evolution process on the transit of electrons through the laccase structure.

The final mutant (R2 variant) was fairly active and stable at concentrations as high as 50% of organic co-solvents, (retaining between 20-30% of its activity in aqueous solution). Moreover, this variant showed promiscuity for different organic co-solvents (DMSO, DMF, DMA, acetonitrile, acetone, ethanol, methanol). Significantly, the spectro-electrochemical features of the enzyme were slightly modified (*i.e.* E^{o}_{T1} and copper geometries), although these changes did not exert any important differences in either its kinetics or stability. R2 harboured four mutations in the mature protein (E182K, S280N, L429N and N552H), as well as two mutations at the C-terminal tail (G8D and E9K). The mutations at the C-terminal tail seemed to affect protein folding, whereas some mutations at the surface of the mature laccase established new interactions, either through salt bridges or hydrogen bonds. These novel interactions were reflected in the structural reinforcement of the regions amenable to denaturation under harsh conditions.

Fungal laccases are also very attractive for application in processes where basic pH is a requisite (*e.g.* cofactor regeneration or pulp biobleaching). Interestingly, the pH profile of MtL shifted progressively toward more alkaline pH in the course of evolution for secretion in yeast and stabilization in organic co-solvents. This unexpected side-effect can be attributed to the fact that laccase mutant libraries were screened at pH 5.0. Therefore, we decided to use the R2 variant as the starting point for evolution towards activity at alkaline pH [58]. R2 was subjected to five further rounds of directed evolution using *ep*PCR combined with different *in vivo* and *in vitro* DNA recombination methods for library creation, and establishing the ratio of activity at pH 8.0 to that at pH 5.0 as selection factor. After screening over 12,000 clones, the final mutant IG-88 showed pH activity profiles both for ABTS and DMP highly shifted towards basic pH (keeping ~90%

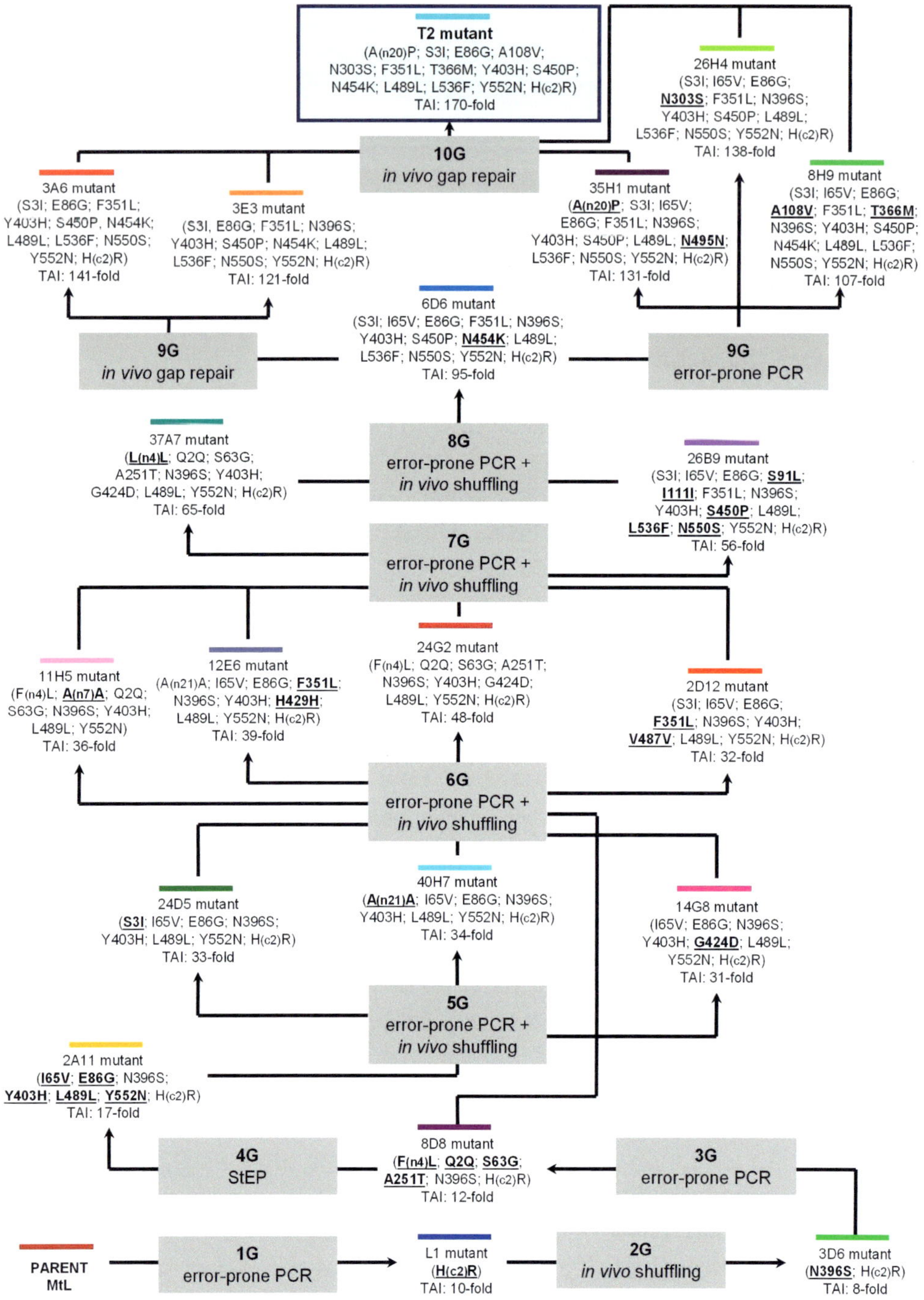

Fig. (2). Artificial evolution pathway for MtL in yeast. TAI: total activity improvement over MtL parent type.

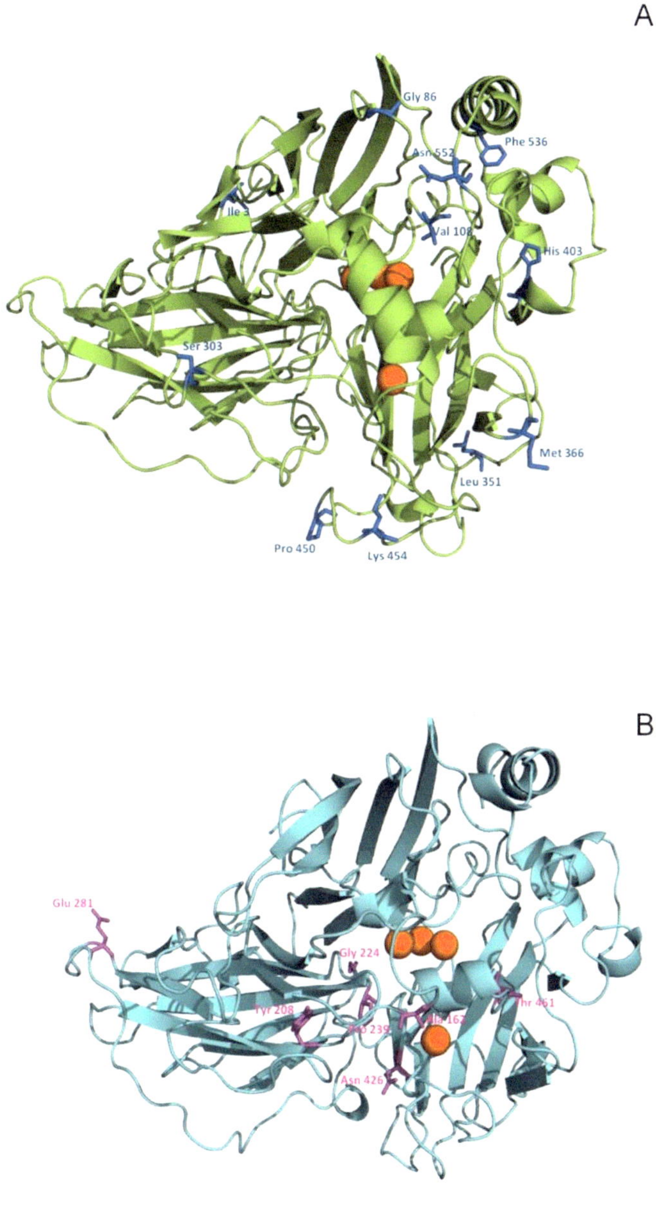

Fig. (3). Amino acid substitutions accumulated in evolved laccases for functional expression in *S. cerevisiae*. **A**) MtLT2-variant; **B**) HRPLs PM1 (OB-1 variant). Orange spheres represent copper atoms. Amino acid substitutions found in mature MtLT2 and OB-1 are highlighted with stick structures.

of its activity at pH 4.0-6.0, 50% at pH 7.0, and even some activity detectable at pH 8.0). IG-88 harboured two additional mutations with respect to the parent type R2 (N109S and D530E), being one of them in a highly conserved position (Asn109) previously reported to be involved in retaining the overall geometry of the trinuclear Cu site in *Melanocarpus albomyces* laccase [59].

In summary, a total of 20 rounds of laboratory evolution were performed on the MtL, giving rise to the accumulation of 16 amino acid mutations in the mature laccase and improved properties in terms of secretion, activity, stability and pH profiles, (Table **1**).

Table 1. Laboratory evolution history of MtL.

Goal of the directed evolution study	**Number of rounds**	**Number of screened clones**	**Final variant**	**Amino acid substitutions in the final variant[1,2,3]**	**Improved features in the final variant**	**Reference**
Functional expression in *S. cerevisiae*	10	20,000	T2	A(sp20)P, S3I, E86G, A108V, N303S, F351L, T366M, Y403H, S450P, N454K, L536F, Y552N, H(c2)R	Expression levels in *S. cerevisiae* of 18 mg/L	Bulter *et al.*, 2003
Organic solvent tolerance	5	13,000	R2	A(sp20)P, S3I, E86G, A108V, <u>E182K</u>, <u>S280N</u>, N303S, F351L, T366M, Y403H, <u>L429N</u>, S450P, N454K, L536F, <u>N552H</u>, H(c2)R, <u>G(c8)D</u>, <u>E(c9)K</u>	High activity and stability at high concentrations of organic solvents	Zumárraga *et al.*, 2007
Activity at alkaline pH	5	12,000	IG-88	A(sp20)P, S3I, E86G, A108V, **N109S**, <u>E182K</u>, <u>S280N</u>, N303S, F351L, T366M, Y403H, <u>L429N</u>, S450P, N454K, **D530E**, L536F, <u>N552H</u>, H(c2)R, <u>G(c8)D</u>, <u>E(c9)K</u>	pH activity profiles significantly shifted towards basic pH	Torres-Salas *et al.*, 2013

[1]Amino acid substitutions responsible for organic solvent tolerance are underlined, while those involved in activity at alkaline pH are highlighted in bold.
[2]Amino acid substitutions in the signal peptide are indicated by sp.
[3]Amino acid substitutions in the C-terminus are indicated by c.

2.3. Directed Evolution of HRPLs

The past successes with MtL evolution cannot easily be translated to their HRPL counterparts, in part because MtL is an ascomycete laccase that facilitates its functional expression in *S. cerevisiae*. Several directed evolution experiments have been attempted by *ep*PCR using HRPLs from *Pleurotus ostreatus* [60-62]. The results confirmed a general improvement in the total activity but poor secretion limits their practical engineering for other purposes.

We described for the first time how to engineer HRPLs that can be expressed strongly by *S. cerevisiae*, enhancing their activities and thermostabilities [24]. Two different HRPLs were used to achieve this goal, the laccase from basidiomycete PM1 (PM1L) [24] and the laccase from *Pycnoporus cinnabarinus* (PcL) [26]. Several fusion proteins were tested by replacing the native signal sequences with others used successfully during heterologous expression in yeast. The best result was obtained with the α-factor prepro-leader and accordingly, the α-PM1L and α-PcL fusion proteins could then be subjected to the corresponding artificial evolution pathways. Interestingly, the joint evolution of the α-factor prepro-leader plus the HRPL allowed us to adjust each to the different protease specificities of the heterologous host throughout its transit from the endoplasmic reticulum to the Golgi compartment. For α-PM1L, eight rounds of evolution were carried out in combination with rational approaches, Fig. (**4**). After screening over 50,000 clones generated by random mutagenesis, *in vivo* DNA shuffling, IvAM and site directed mutagenesis, the total laccase activity (which is the product of its secretion and kinetics) was enhanced up to 34,000-fold, the largest improvement ever reported for this kind of system [24]. We attribute such a large improvement to the joint evolution of the prepro-leader and the laccase, which meant that a synergistic effect was produced during its exportation by yeast. The ultimate variant obtained through this evolutionary process (the OB-1 mutant) was readily secreted by *S. cerevisiae* (up to 8 mg/L) in a soluble, very active and very stable form, particularly with respect to temperature (with a T_{50} value of 73ºC), pH and co-solvents, Fig. (**3B**). For α-PcL, six rounds of evolution coupled to a high-throughput (HTS) assay based on the oxidation of natural a synthetic mediators were performed [26]. The final variant of this study (the 3PO mutant) showed 8,000-fold improved activity over the parent type α-PcL. Mutations located in the α-factor prepro-leader were responsible for a 40-fold enhancement in secretion by *S. cerevisiae* (~2 mg/L), while beneficial mutations in the mature protein led to ~14-fold increase in the k_{cat} for ABTS.

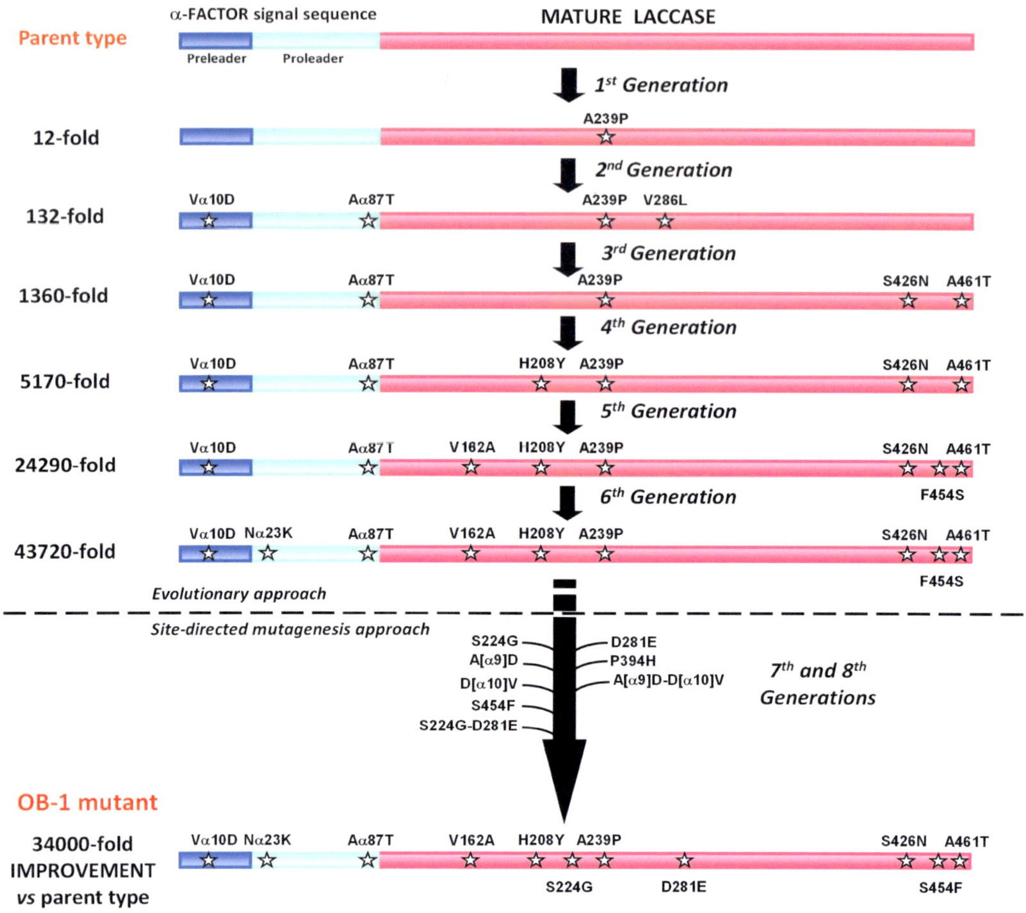

Fig. (4). Combination of directed evolution and rational approaches for the engineering of HRPLs.

Among the strategies engineered to evolve HRPLs, the mutational exchange and the recovery of beneficial mutants should be highlighted. With mutational exchange, several mutations found in both evolutionary programmes (for the α-PM1 and the α-PcL) were switched from one system to another, taking advantage of their close protein sequence identity (above 75%). Interestingly, some mutations found in the hydrophobic core of the α-factor pre-leader were valuable in both systems, opening the possibility of evolving the α-factor prepro-leader as a universal signal peptide for the heterologous expression of fungal laccases in yeast. With mutational recovery, some beneficial mutations ruled out by the yeast recombination apparatus could be recovered by site-directed mutagenesis of OB-1 having mapped them first in the family tree of the whole evolution experiment. Last but not least, we paid special attention to protein stability since the overall philosophy of this work was to create a scaffold on which new functions could be developed and that was sufficiently stable to tolerate the introduction of a new set of beneficial mutations. Hence, the drops in stability produced during evolution by the accumulation of some beneficial but destabilizing mutations could be recovered by rational approaches, coupled with the screening of mutant libraries for thermostability, Fig. (**5**).

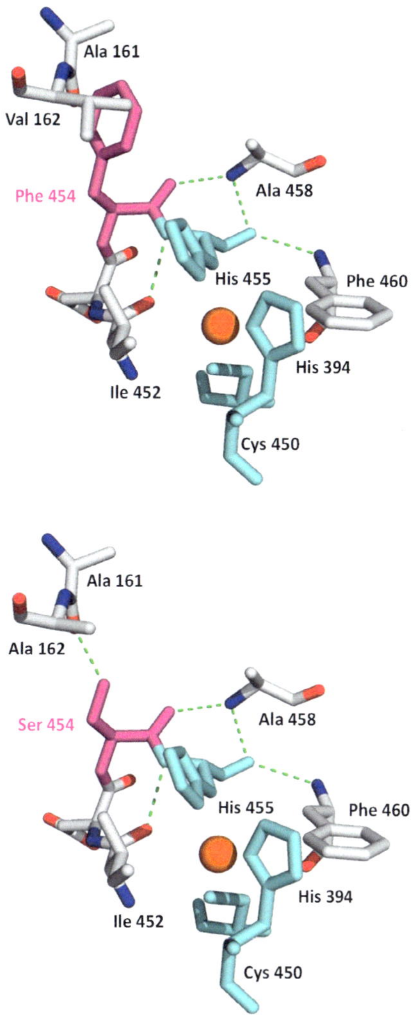

Fig. (5). Rational approach for thermostability during the evolution of PM1 laccase. Mutation F454S discovered in the 5th round enhanced the activity but at the cost of reducing the thermal stability (a hydrogen bond formed with Ala161 may affect the distance coordination between His455 and the T1 Cu). The reverted variant S454F completely recovered its stability which allowed the introduction of a new set of beneficial mutations.

Going one step further, the synergistic use of directed evolution, site-directed mutagenesis and semi-rational approaches has allowed HRPLs to be designed so that they can work in non-natural environments like blood or ionic liquids. The secretion variant of the PM1L evolution (the OB-1 mutant) was selected as parent type to design a laccase able to work in human blood [50]. Like other HRPLs, the OB-1 variant was strongly inhibited by low concentration of chloride ions and was not active at neutral/basic pH. These features precluded to use OB-1 in human blood (with 150 mM NaCl and a pH around 7.4). To evolved OB-1 for activity in blood, a HTS assay based on a surrogate blood (mimicking the biochemical composition of human blood but lacking coagulation factors and blood cells) was prepared. OB-1 was subjected to four rounds of directed evolution together with rational design and saturation mutagenesis. The best variant of the project (the ChU-B mutant) showed a 40,000-fold increase in total activity with respect to the parent type OB-1. The biochemical breakdown of its properties resulted in a notable pH activity profile shifted to the basic side along with the highest tolerance to chloride reported to date for HRPLs, with an increase in the I_{50} for Cl^- from 176 mM to 1025 mM (I_{50} was the concentration of Cl^- at which the laccase keeps 50% of its activity). The ChU-B mutant showed two new mutations in the proximity of the T1 Cu site (F396I and F454E) that gave rise to activity in blood but at the cost of jeopardizing thermostability. More recently, this blood tolerant laccase was immobilized onto chemically modified electrodes, becoming the first laccase capable of catalysing the electro-oxidation of water to molecular oxygen as this process is thermodynamically favoured at neutral pHs [63]. In addition, the ChU-B variant was successfully incorporated into a self-powered biodevice for carbohydrate and oxygen monitoring [64].

Laccase Lcc2 from *Trametes versicolor* was subjected to two rounds of random mutagenesis and screening towards enhanced ionic liquid resistance [65]. The final variant of this study (the M3 mutant) accumulated two mutations in the mature protein (F265S and A318V), which were responsible for an increase of 4.5 times in activity in the presence of 15% (v/v) of the ionic liquid 1-ethyl-3-methylimidazolium ethyl sulfate. The M3 variant represents a good candidate to be used as biocatalyst for lignin degradation in homogeneous ionic solutions and therefore, for second-generation biofuel development.

Last but not least, there are also a few studies in which basidiomycete laccases were evolved for secretion in yeast species other than *S. cerevisiae*. Laccase from *Fomes lignosus* was functionally expressed in *P. pastoris* by constructing mutant libraries through ethyl methane sulfonate-based random mutagenesis [66]. The best laccase mutant (the PPM5 variant) harboured four mutations that gave rise to a 3.7-fold enhancement in expression with respect to the wild-type (144 mg/L), as well as to a 1.4-fold higher k_{cat} for ABTS. Two of the hydrophilic mutations were located close to the water channel, which might help water molecules to get out of the trinuclear Cu cluster, thereby explaining the increase on the k_{cat}. Laccase from *Cyathus bulleri* was engineered also in *P. pastoris* but with the aim of decolorizing the phthalocyanine dye reactive blue 21 (RB21) [67]. The C-terminus of the laccase gene was subjected to *ep*PCR and the laccase library was pre-screened in ABTS-containing agar plates. The highly ABTS-oxidizing clones were tested for decolorization of RB1 in microtiter plates. Three variants (Lcc-35, Lcc-61 and Lcc-62) were found to show around 78% to 95% decolorization in 30 min using ABTS as redox mediator. Structural analysis of the modelled structure revealed mutations both far and close to T1 Cu site that enhanced catalytic efficiency for ABTS and, in turn, the oxidation rate of RB21. On the other hand, the lcc1 gene coding for *T. versicolor* laccase was cloned into the genome of *Yarrowia lipolytica* using single and multiple integration sites [68]. After checking for successful secretion into the culture media, *ep*PCR followed by site-directed mutagenesis was performed over the expression cassette with a single integration site. As a result, a laccase variant (the rM-4A mutant) with a 5.8-fold improved activity over wild-type and better catalytic efficiencies towards ABTS and DMP was identified.

CONCLUSIONS: WHAT HAVE WE LEARNT?

Over the years, *S. cerevisiae* has helped us to engineer strategies for the directed evolution of fungal laccases. The low redox potential laccase MtL was the first successful example [46] and since then, we have designed specific approaches for the laboratory evolution of HRPLs that combine *in vivo* and *in vitro* tools with rational approaches [24, 26, 50]. The evolved fungal laccases mentioned in this review constitute unique platforms for further protein engineering through directed evolution, principally aimed at generating enzymes that can be used in attractive biotechnological applications. Indeed, we are now in a position to tailor HRPLs for different purposes, ranging from the increase of the redox potential at the T1 Cu to the construction of artificial ligninolytic secretome in yeast [2, 11, 69]. There are still some hurdles to overcome and some questions that should be answered: why is the functional expression of laccases in *S. cerevisiae* so tremendously difficult? Would it be possible to evolve high-redox potential laccases in bacteria? Is it feasible to enhance the laccase redox potential beyond natural limits by directed evolution or by rational means without disturbing its stability?... and more significantly, would that improvement necessarily mean better activity? We hope that in the near future, new HRPLs engineered by directed evolution and rational approaches can affront the attractive challenges presented by traditional and modern biocatalysis.

CONFLICT OF INTEREST

The author confirms that he has no conflict of interest to declare for this publication.

ACKNOWDLEDGEMENTS

This work was supported by the European Commission projects Indox-FP7-KBBE-2013-7-613549 and Cost-Action CM1303-Systems Biocatalysis, and the National Projects Dewry [BIO201343407-R] and Cambios [RTC-2014-1777-3].

DISCLOSURE

This is an updated chapter of the earlier review article [by Maté, D., García-Ruiz, E., Camarero, S., & Alcalde, M. entitled "Directed evolution of fungal laccases". *Current Genomics*, 2011; 12(2), 113], with additional text and references.

ABBREVIATIONS

ABTS	=	2,2-azino-bis[3-ethylbenzthizoline-6-sulfonic acid]
CLERY	=	combinatorial libraries enhanced by recombination in yeast
CFU	=	colony forming units
DMA	=	dimethylacetamide
DMF	=	dimethylformamide
DMP	=	2,6-dimethoxyphenol
DMSO	=	dimethyl sulfoxide
E^{o}_{T1}	=	redox potential at the T1 Cu site
*ep*PCR	=	error-prone PCR
GAP	=	glyceraldehyde 3-phosphate dehydrogenase gene promoter from *P. pastoris*

HRPLs	=	high-redox potential laccases
HTS	=	high-throughput screening
IvAM	=	*in vivo* assembly of mutant libraries with different mutational spectra
IVOE	=	*in vivo* overlap extension
MORPHING	=	mutagenic organized recombination process by homologous *in vivo* grouping
MtL	=	*Myceliophthora thermophila l*accase
NHE	=	normal hydrogen electrode
PAH	=	polycyclic aromatic hydrocarbon
PcL	=	*Pycnoporus cinnabarinus* laccase
PARS1	=	pichia autonomous replication sequence 1
PM1L	=	laccase from basidiomycete PM1
RB21	=	reactive blue 21
StEP	=	staggered extension process
T_{50}	=	temperature at which the enzyme retains 50% of its activity after incubation for 10 min
α-PcL	=	fusion gene of the α-factor prepro-leader and mature PcL
α-PM1	=	fusion gene of the α-factor prepro-leader and mature PM1 laccase

REFERENCES

[1] Gianfreda L, Xu F, Bollag J. Laccases: a useful group of oxidoreductive enzymes. Biorem J 1999, 3(1): 1-25.

[2] Rodgers CJ, Blandford CF, Giddens SR, Skamnioti P, Armstrong FA, Gurr SJ. Designer laccases: a vogue for high-potential fungal enzymes?. Trends Biotechnol 2010; 28(2): 63-72.

[3] Alcalde M. Laccases: biological functions, molecular structure and industrial applications. In: Polaina J, MacCabe AP, Eds. Industrial enzymes: structure, functions and applications. New York: Springer 2007; pp. 459-74.
[4] Alcalde M, Ferrer M, Plou FJ, Ballesteros A. Environmental biocatalysis: from remediation with enzymes to novel green processes. Trends Biotechnol 2006; 24(6): 281-87.
[5] Kunamneni A, Ballesteros A, Plou FJ, Alcalde M. Fungal laccase: a versatile enzyme for biotechnological applications. In: Méndez-Vilas A, Ed. Communicating current research and educational topics and trends in applied microbioloy. Spain: Formatex Publishers 2007; pp. 233-45.
[6] Giardina P, Faraco V, Pezzella C, Piscitelli A, Vanhulle S, Sannia G. Laccases: a never-ending story. Cell Mol Life Sci 2010; 67(3): 369-85.
[7] Solomon EI, Sundaram UM, Machonkin TE. Multicopper oxidases and oxygenases. Chem Rev 1996; 96(7): 2563-605.
[8] Mayer AM, Staples RC. Laccase: new functions for an old enzyme. Phytochemistry 2002; 60(6): 551-65.
[9] Claus H. Laccases: structure, reactions, distribution. Micron 2004; 35(1-2): 93-6.
[10] Cañas A, Camarero S. Laccases and their natural mediators: biotechnological tools for sustainable eco-friendly processes. Biotechnol Adv 2010; 28(6): 694-705.
[11] Mate DM, Alcalde M. Laccase engineering: from rational design to directed evolution. Biotechnol Adv 2015; 33(1): 25-40.
[12] Gupta N, Farinas ET. Narrowing Laccase Substrate Specificity Using Active Site Saturation Mutagenesis. Comb Chem High Throughput Screen 2009; 12(3): 269-74.
[13] Gupta N, Lee FS, Farinas ET. Laboratory evolution of laccase for substrate specificity. J Mol Catal B: Enzym 2010; 62(3-4): 230-4.
[14] Gupta N, Farinas ET. Directed evolution of CotA laccase for increased substrate specificity using *Bacillus subtilis* spores. Protein Eng Des Sel 2010; 23(8): 679-82.
[15] Koschorreck K, Schmid R, Urlacher V. Improving the functional expression of a *Bacillus licheniformis* laccase by random and site-directed mutagenesis. BMC Biotechnol 2009; 9: 12.
[16] Liu YH, Ye M, Lu Y, Zhang X, Li G. Improving the decolorization for textile dyes of a metagenome-derived alkaline laccase by directed evolution. Appl Microbiol Biotechnol 2011; 91(3): 667-75.
[17] Xu F. Applications of oxidoreductases: recent progress. Ind Biotechnol 2005; 1(1): 38-50.
[18] Kunamneni A, Plou FJ, Ballesteros A, Alcalde M. Laccase and their applications: a patent review. Recent Pat Biotechnol 2008; 2(1): 10-24.
[19] Kunamneni A, Camarero S, García C, Plou FJ, Ballesteros A, Alcalde M. Engineering and applications of fungal laccases for organic synthesis. Microb Cell Fact 2008; 7: 32.
[20] Riva S. Laccase: blue enzymes for green chemistry. Trends Biotechnol 2006; 24(5): 219-26.
[21] Palmer AE, Szilagyi RK, Cherry JR, Jones A, Xu F, Solomon EI. Spectroscopic characterization of the Leu513His variant of fungal laccase: Effect of increased axial ligand interaction on the geometric and electronic structure of the type 1 Cu site. Inorg Chem 2003; 42(13): 4006-17.
[22] Xu F, Palmer AE, Yaver DS, *et al.* Targeted mutations in a *Trametes villosa* laccase. Axial perturbations of the T1 copper. J Biol Chem 1999; 274(18): 12372-5.
[23] Xu F, Berka RM, Wahleithner JA, *et al.* Site-directed mutations in fungal laccase: effect on redox potential, activity and pH profile. Biochem J 1998; 334: 63-70.
[24] Maté D, García-Burgos C, García-Ruiz E, Ballesteros AO, Camarero S, Alcalde M. Laboratory evolution of high-redox potential laccases. Chem Biol 2010; 17(9): 1030-41.
[25] García-Ruiz E, Maté D, Ballesteros A, Martínez AT, Alcalde M. Evolving thermostability in mutant libraries of ligninolytic oxidoreductases expressed in yeast. Microb Cell Fact 2010; 9: 17.
[26] Camarero S, Pardo I, Cañas AI, *et al.* Engineering platforms for directed evolution of laccase from *Pycnoporus cinnabarinus*. Appl Environ Microbiol 2012; 78(5): 1370-84.
[27] Tracewell CA, Arnold FH. Directed enzyme evolution: climbing fitness peaks one amino acid at a time. Curr Opin Chem Biol 2009; 13(1): 3-9.
[28] Bloom JD, Arnold FH. In the light of directed evolution: pathways of adaptive protein evolution. Proc Natl Acad Sci USA 2009; 106: 9995-10000.
[29] Romero PA, Arnold FH. Exploring protein fitness landscapes by directed evolution. Nat Rev Mol Cell Biol 2009; 10: 866-76.

[30] Alcalde M, Bulter T. Colorimetric assays for screening laccases. In: Arnod FH, Georgiou G, Eds. Directed enzyme evolution. Screening and selection methods. Totowa: Humana Press 2003; pp. 193-201.

[31] Alcalde M, Bulter T, Zumárraga M, *et al.* Screening mutant libraries of fungal laccases in the presence of organic solvents. J Biomol Screen 2005; 10(6): 624-31.

[32] Alcalde M, Bulter T, Arnold FH. Colorimetric assays for biodegradation of polycyclic aromatic hydrocarbons by fungal laccases. J Biomol Screen 2002; 7(6): 547-53.

[33] Zumárraga M, Plou FJ, García H, Ballesteros A, Alcalde M. Biorremediation of polycyclic aromatic hydrocarbons by fungal laccases engineered by directed evolution. Biocatal Biotransfor 2007; 25(2-4): 219-28.

[34] Pardo I, Chanagá X, Vicente AI, Alcalde M, Camarero S. New colorimetric screening assays for the directed evolution of fungal laccases to improve the conversion of plant biomass. BMC Biotechnol 2013; 13: 90.

[35] Alcalde M. Mutagenesis protocols in *Saccharomyces cerevisiae* by *in vivo* overlap extension. In: Bramman J, Ed. *In vitro* mutagenesis protocols. Totowa: Humana Press 2010; pp. 3-14.

[36] Zumárraga M, Domínguez CV, Camarero S, *et al.* Combinatorial saturation mutagenesis of the *Myceliophthora thermophila* laccase T2 mutant: the connection between the C-terminal plug and the conserved $_{509}$VSG$_{511}$ tripeptide. Comb Chem High Throughput Screen 2008; 11(10): 807-16.

[37] Arnold FH. Fancy footwork in the sequence space shuffle. Nature Biotechnol 2006; 24(3): 328-30.

[38] Gonzalez-Perez D, Garcia-Ruiz E, Alcalde M. *Saccharomyces cerevisiae* in directed evolution: An efficient tool to improve enzymes. Bioengineered 2012; 3(3): 172-7.

[39] Pourmir A, Johannes TW. Directed evolution: selection of the host microorganism. Comput Struct Biotechnol J 2012; 2: e201209012.

[40] Lee CC, Williams TG, Wong DWS, Robertson GH. An episomal expression vector for screening mutant gene libraries in *Pichia pastoris*. Plasmid 2005; 54(1): 80-5.

[41] Sandström AG, Engström K, Nyhlén J, Kasrayan A, Bäckvall J-E. Directed evolution of *Candida antarctica* lipase A using an episomaly replicating yeast plasmid. Protein Eng Des Sel 2009; 22(7): 413-20.

[42] Liu Z, Pscheidt B, Avi M, *et al.* Laboratory evolved biocatalysts for stereoselective syntheses of substituted benzaldehyde cyanohydrins. ChemBioChem 2008; 9(1): 58-61.

[43] Zumárraga M, Camarero S, Shleev S, *et al.* Altering the laccase functionality by *in vivo* assembly of mutant libraries with different mutational spectra. Proteins 2008; 71(1): 250-60.

[44] Alcalde M, Zumárraga M, Polaina J, Ballesteros A, Plou FJ. Combinatorial saturation mutagenesis by *in vivo* overlap extension for the engineering of fungal laccases. Comb Chem High Throughput Screen 2006; 9(10): 719-27.

[45] Cherry JR, Lamsa MH, Schneider P, *et al.* Directed evolution of a fungal peroxidase. Nat Biotechnol 1999; 17(4): 379-84.

[46] Bulter T, Alcalde M, Sieber V, Meinhold P, Schlachtbauer C, Arnold FH. Functional expression of a fungal laccase in *Saccharomyces cerevisiae* by directed evolution. Appl Environ Microbiol 2003; 69(2): 987-95.

[47] Okkels JS. *In vivo* gene shuffling in yeast - a fast and easy method for directed evolution of enzymes In: Svendsen A, Ed. Enzyme functionality: design, engineering and screening. New York: CRC Press 2004; pp. 413-24.

[48] Cusano AM, Mekmouche Y, Meglecz E, Tron T. Plasticity of laccase generated by homeologous recombination in yeast. FEBS Lett 2009; 276(19): 5471-80.

[49] Zumárraga M, Bulter T, Shleev S, *et al. In vitro* evolution of a fungal laccase in high concentrations of organic cosolvents. Chem Biol 2007; 14(9): 1052-64.

[50] Mate DM, Gonzalez-Perez D, Falk M, *et al.* Blood tolerant laccases by directed evolution. Chem Biol 2013; 20(2): 223-31.

[51] Gonzalez-Perez D, Molina-Espeja P, Garcia-Ruiz E, Alcalde M. Mutagenic organized recombination process by homologous *in vivo* grouping (MORPHING) for directed enzyme evolution. PLOS One 2014; 9(3): e90919.

[52] Gonzalez-Perez D, Garcia-Ruiz E, Ruiz-Dueñas FJ, Martinez AT, Alcalde M. Structural determinants of oxidative stabilization in an evolved versatile peroxidase. ACS Catal 2014; 4(11): 3891-901.

[53] Molina-Espeja P, Garcia-Ruiz E, Gonzalez-Perez D, Ullrich R, Hofrichter M, Alcalde M. Directed evolution of unspecific peroxygenase from *Agrocybe aegerita.* Appl Environ Microbiol 2014; 80(11): 3496-507.
[54] Viña-Gonzalez J, Gonzalez-Perez D, Ferreira P, Martinez AT, Alcalde M. Focused directed evolution of aryl-alcohol oxidase in yeast using chimeric signal peptides. Appl Environ Microbiol 2015; 81(18): 6451-62.
[55] Abécassis V, Pompon D, Truan G. High efficiency family shuffling based on multi-step PCR and *in vivo* DNA recombination in yeast: statistical and functional analysis of a combinatorial library between human cytochrome P450 1A1 and 1A2. Nucleic Acids Res 2000; 28(20): e88.
[56] Pardo I, Vicente AI, Mate DM, Alcalde M, Camarero, S. Development of chimeric laccases by directed evolution. Biotechnol Bioeng 2012; 109(12): 2978-86.
[57] Garcia-Ruiz E, Gonzalez-Perez D, Ruiz-Dueñas FJ, Martinez AT, Alcalde M. Directed evolution of a temperature, peroxide and alkaline pH tolerant versatile peroxidase. Biochem J 2012; 441(1): 487-98.
[58] Torres-Salas P, Mate DM, Ghazi I, Plou FJ, Ballesteros AO, Alcalde M. Widening the pH activity profile of a fungal laccase by directed evolution. ChemBioChem 2013; 14(8): 934-7.
[59] Andberg M, Hakulinen N, Auer S, *et al.* Essential role of the C-terminus in *Melanocarpus albomyces* laccase for enzyme production, catalytic properties and structure. FEBS J 2009; 276(21): 6285-300.
[60] Festa G, Autore F, Fraternali F, Giardina P, Sannia G. Development of new laccases by directed evolution: functional and computational analyses. Proteins 2008; 72(1): 25-34.
[61] Miele A, Giardina P, Sannia G, Faraco V. Random mutants of a *Pleurotus ostreatus* laccase as new biocatalysts for industrial effluents bioremediation. J Appl Microbiol 2010; 108(3): 998-1006.
[62] Miele A, Giardina P, Notomista E, Piscitelli A, Sannia G, Faraco V. A semi-rational approach to engineering laccase enzymes. Mol Biotechnol 2010; 46(2): 149-56.
[63] Pita M, Mate DM, Gonzalez-Perez D, *et al.* Bioelectrochemical oxidation of water. J Am Chem Soc 2014; 136(16): 5892-5.
[64] Falk M, Alcalde M, Bartlett PN, *et al.* Self-powered wireless carbohydrate/oxygen sensitive biodevice based on radio signal transmission. PLOS One 2014; 9(10): e109104.
[65] Liu H, Zhu L, Bocola M, Chen N, Spiess AC, Schwaneberg U. Directed laccase evolution for improved ionic liquid resistance. Green Chem 2013; 15: 1348-55.
[66] Hu MR, Chao YP, Zhang GQ, Yang XQ, Xue ZQ, Qian SJ. Molecular evolution of *Fome lignosus* laccase by ethyl methane sulfonate-based random mutagenesis *in vitro*. Biomol Eng 2007; 24(6): 619-24.
[67] Kenzom T, Srivastava P, Mishra S. Structural insights into 2,2'-azino-bis(3-ethylbenzothiazoline-6-sulfonic acid) (ABTS)-mediated degradation of reactive blue 21 by engineered *Cyathus bulleri* laccase and characterization of degradation products. Appl Environ Microbiol 2014; 80(24): 7484-95.
[68] Theerachat M, Emond S, Cambon E, *et al.* Engineering and production of laccase from *Trametes versicolor* in the yeast *Yarrowia lipolytica*. Bioresource Technol 2012; 125: 267-74.
[69] Gonzalez-Perez D, Alcalde M. Assembly of evolved ligninolytic genes in *Saccharomyces cerevisiae.* Bioengineered 2014; 5(4): 254-63.

CHAPTER 6

The MADS-box Genes Involved in Orchid Flower Development

Serena Aceto* **and Luciano Gaudio**

Department of Biology, University of Naples Federico II, Napoli, Italy

Abstract: Since the advent of Darwinian genetics, there has been much interest in the evolutionary origin of the Orchidaceae, one of the most species-rich angiosperm families. Orchids have highly diversified and specialized flowers, and some species exhibit an uncoupled rate of morphological and molecular evolution. Recently, these peculiar characteristics have enhanced the study of the orchid MADS-box genes involved in flower development. This large gene family encodes transcription factors that constitute the main regulatory network driving the formation of flower organs. Recent analyses have highlighted the role of the MADS-box genes in orchids and shown that different evolutionary forces act on the coding and non-coding regions of these genes. The most widely accepted theory proposed to explain the evolution of the orchid perianth is the "orchid code", which posits that the orchid floral organs became diversified through a series of duplications and mutations of the MADS-box genes, followed by functional diversification.

Keywords: ABCDE model, adaptation, column, consensus sequence, development, evolution, flower, inflorescence, labellum, MADS-box, Orchidaceae, orchid code, ovary, phylogeny, quartet model, regulation, resupination, tepals, transcription, zygomorphy.

INTRODUCTION

Among the angiosperms, the Orchidaceae is one of the families with the highest number of species and is characterized by flowers with extremely divergent morphologies and specialized organs. Orchid species have colonized widely different habitats. Epiphytism, specialized pollination strategies [1], natural selection and genetic drift [2] are considered the main causes of their evolutionary success.

The orchid flower is known for its intriguing beauty and mystery, and despite great morphological variability among the different species, its organization can

***Corresponding author Serena Aceto:** Department of Biology, University of Naples Federico II, *via* Mezzocannone 8, 80134 Napoli, Italy; Tel: +390812535190; Fax: +390812535035; E-mail: serena.aceto@unina.it

Atta-ur-Rahman (Ed.)

be traced to a common scheme. The orchid flower is zygomorphic (bilaterally symmetrical) and includes three outer tepals (sepals), two lateral inner tepals (petals) and a highly diversified median inner tepal (lip or labellum) (Fig. **1**). The reproductive structure of the orchid flower is the column, a fusion of male (stamen/anther) and female (pistil/stigma) tissues. The pollen grains (pollinia) are located at the top of the column, whereas the ovary, whose development is triggered by pollination, is placed at the base of the column [3-5]. The outer tepals can have different shapes and pigmentations, can be similar to the lateral inner tepals, or can form nectar spurs. Generally considered the most distinctive structure of the orchid perianth, the lip has a peculiar shape and color pattern and is sometimes adorned with calli, spurs and glands. The lip is homologous to the adaxial tepal of other monocots and, therefore, should be the uppermost tepal; however, during the development of the orchid flower, resupination takes place: the 180° rotation in floral orientation that shifts the lip to the lowest tepal position. The orientation and position opposite to the fertile anther of the lip after resupination suggest that the morphological diversifications in lip shape and color are the result of specific adaptations to pollinators [6].

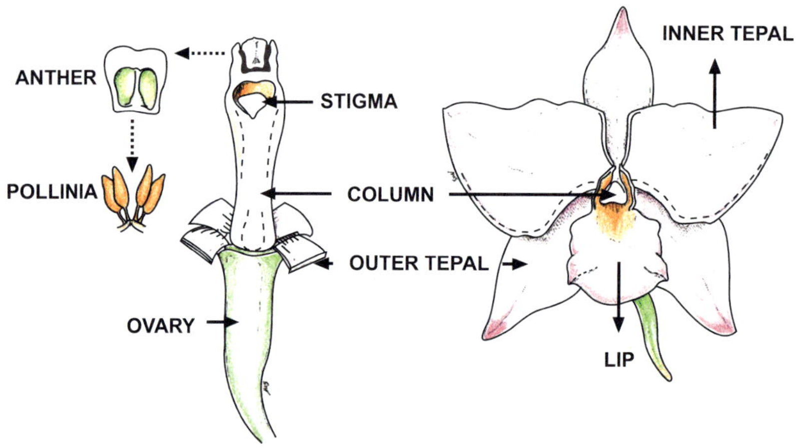

Figure 1: Scheme of an orchid flower.

Before the widespread use of molecular techniques, the phylogeny of Orchidaceae was based on a restricted number of morphological characteristics. Some of these characteristics, *e.g.*, the floral structures involved in the interactions between orchids and pollinators, are adaptive in nature and therefore may not be useful for reconstructing the actual phylogeny of orchids. In fact, a number of conflicting taxonomic and phylogenetic reconstructions of the Orchidaceae have been proposed [7]. In more recent years, the increasing utilization of molecular markers and morphological traits has deeply expanded and improved the taxonomy and phylogeny of orchids. Based on the currently accepted classification system, the family Orchidaceae includes five subfamilies: Apostasioideae, Cypripedioideae, Epidendroideae, Orchidoideae and Vanilloideae (Fig. **2**). Each subfamily is further divided into a large number of tribes and subtribes [8, 9].

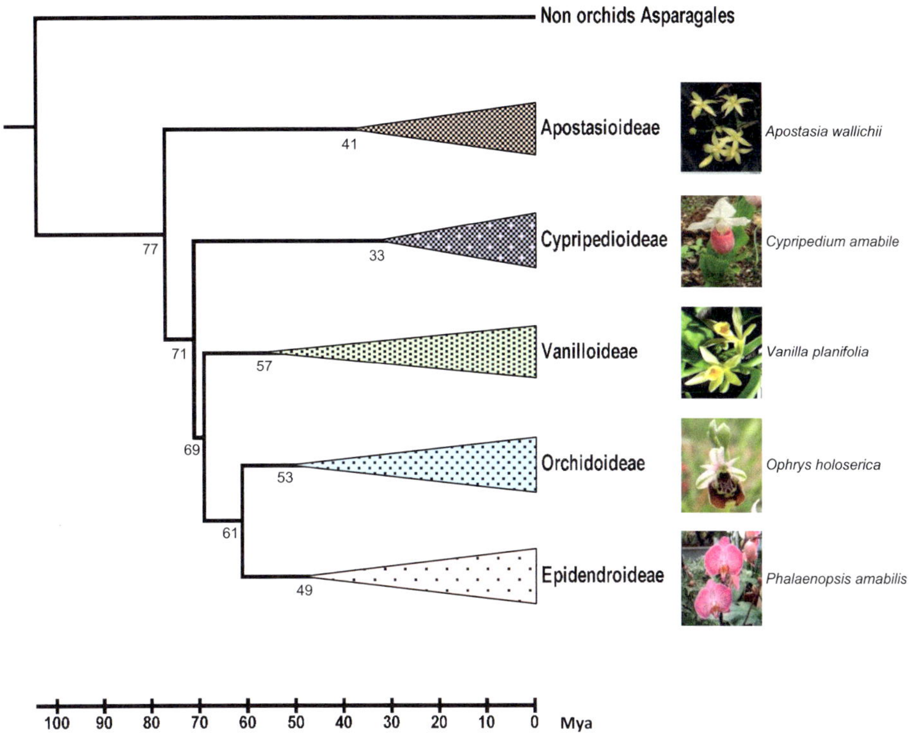

Figure 2: Time-calibrated phylogeny of the Orchidaceae subfamilies, modified from Gustafsson *et al.* [14]. The divergence time is indicated below the branches and is expressed in millions of years ago (Mya). Images of orchid species representative of each subfamily are on the right.

Another critical issue concerns the temporal origin and diversification of orchids. Initially, the origin of the modern orchids was placed between ~26 and ~110 million years ago (Mya) [10-13], until the discovery of a fossil of the extinct stingless bee *Proplebeia dominicana*, dated 15-20 Mya, covered with pollen grains of the orchid species *Meliorchis caribea*. This find restricted the time interval of the orchid family's origin, placing it in the Late Cretaceous, approximately 76-84 Mya [12]. This dating was confirmed by a more recent analysis [14] that included two new orchid fossils classified as *Dendrobium* and *Earina* [15], both of which placed the origin of the most recent common ancestor at ~77 Mya, although the origin of the five orchid subfamilies is approximately 1-8 Mya prior to the previous dating [14] (Fig. **2**). These recent calibration studies [12, 14] used the plastid *matK* and *rbcL* nucleotide sequences to infer the phylogenetic reconstructions, confirming the importance of molecular analyses in the study of the origin and evolution of orchids.

Although Orchidaceae possess many peculiar characteristics, such as highly specialized reproductive strategies, extremely variable flower morphologies, specific ecological strategies and developmental pathways, past molecular studies on orchids were quite limited, particularly when compared with those of other large flowering plant families [16]. Recent advances in next-generation sequencing techniques have greatly increased the amount of molecular data available on orchids. The assembled genome of the orchid *Phalaenopsis equestris* was recently released [17], and a genome-sequencing project for *Phalaenopsis aphrodite* (NCBI BioProject ID 53151) is under development. In addition, an increasing number of *de novo* assembled orchid transcriptomes have been recently published [18-31]. The OrchidBase 2.0 (http://orchidbase.itps.ncku.edu.tw/) and Orchidstra (http://orchidstra.abrc.sinica.edu.tw) are two freely available collections of nucleotide sequences obtained from different RNA-seq projects that provide integrated information on the transcriptomes of various orchid species [32-34]. The establishment of such public resources is important, as it facilitates research into different aspects of orchid biology.

In this chapter, we will review the currently known molecular pathways underlying the development of the flower, the most specialized and diversified orchid structure, with particular regard to the function of the MADS-box genes.

The MADS-box Genes

The name of the MADS-box genes family is taken from the initials of four loci, *MCMI* in *Saccharomyces cerevisiae*, *AG* in *Arabidopsis thaliana*, *DEF* in *Antirrhinum majus* and *SRF* in *Homo sapiens*. These loci encode transcription factors that contain a MADS-box domain, a conserved DNA-binding domain of 56 amino acids [35]. The MADS-box genes originated from subunit A of topoisomerase II [36] and have been found in almost all eukaryotes, although these genes have become greatly expanded only in land plants. The MADS-box genes are divided into two main groups, types I and II, that originate from a gene duplication event prior to the divergence of plants and animals. Type I and II MADS-box genes have different genomic organizations, developmental roles, evolutionary rates and levels of functional redundancy [37]. The type I genes are further divided into three groups, Malpha, Mbeta and Mgamma [38], and participate in the development of female gametophytes, seeds and embryos [39]. The type II MADS-box proteins share the so-called MIKC structure, which includes the highly conserved DNA-binding MADS domain (M) at the amino terminus, a poorly conserved domain (I) and a moderately conserved domain (K) in the middle (both of which are involved in protein–protein interactions and in the formation of coiled-coil structures), and a highly variable carboxyl-terminal (C) region that may function as a transactivation domain [40, 41]. Based on the intron/exon organization of the I domain, the MIKC genes are divided into the $MIKC^C$ and MIKC* genes [42]. Functional studies have revealed that the MIKC* genes are mainly involved in the development of the male gametophyte [43], whereas the $MIKC^C$ genes play many different roles during plant growth and development, including the establishment and maintenance of the floral organs. The $MIKC^C$ genes represent the most studied and best-characterized group of the MADS-box genes and are usually simply referred to as the MIKC genes. Their genomic organization is generally conserved and includes seven introns and eight exons [41, 44-48].

The expression of the MADS-box genes is controlled by regulatory networks integrated in a complex cascade of events [41, 46, 49], sometimes involving highly specialized mechanisms, such as regulation by small RNAs [50] and epigenetic control [51]. The increasing amount of data obtained from genome projects and reverse genetic studies is clarifying in even more detail the mechanisms underlying the evolutionary origin and functional diversification of this complex family of transcription factors [52].

The ABCDE model of flower development describes in a schematic manner the main interactions of the floral homeotic genes from a spatial and functional perspective (Fig. **3A**) [53, 54]. The first proposal of this model was based on an analysis of homeotic mutants of the model species *Arabidopsis thaliana*, whose flowers are comprised of four concentric whorls of floral organs. Excluding the gene *APETALA2* (*AP2*), all the genes involved in the ABCDE model belong to different functional classes of the MADS-box genes. In *Arabidopsis*, the expression of the class A genes (*APETALA1*, *AP1*) regulates sepal development in whorl 1. The synergistic activities of the class A and B genes (*e.g.*, *PISTILLATA*, *PI*, and *APETALA3*, *AP3*) controls the formation of petals in whorl 2. In whorl 3, the expression of the class B genes together with the expression of the class C genes (*e.g.*, *AGAMOUS*, *AG*) determines stamen development. The expression of the class C genes alone drives the formation of the carpel in whorl 4. The class D genes (*e.g.*, *SEEDSTICK*, *STK,* and *SHATTERPROOF*, *SHP*) modulate the identity of the ovule, and the class E genes (*e.g.*, *SEPALLATA*, *SEP*) are involved in the formation of all the floral organs.

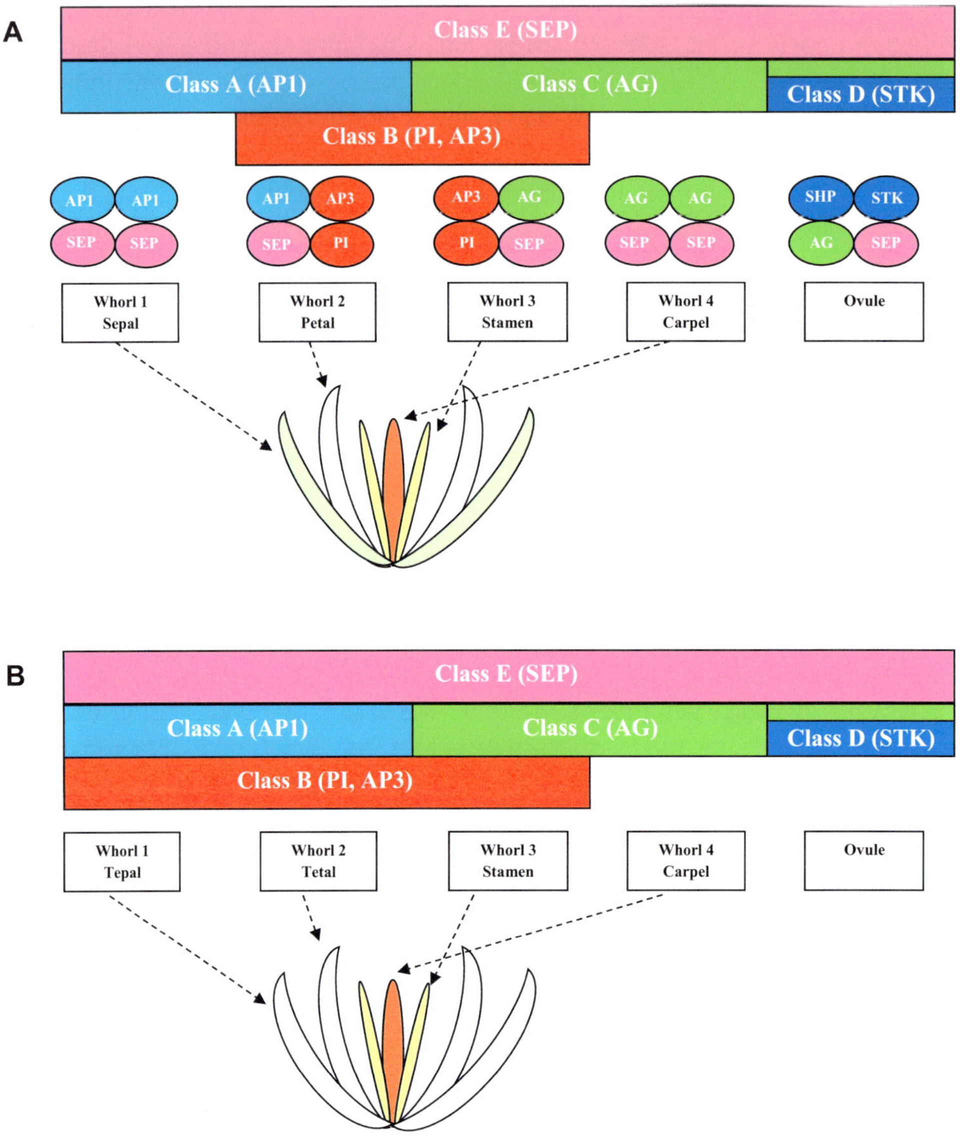

Figure 3: Schemes of the ABCDE and quartet models (**A**) and of the modified ABCDE model of floral development (**B**).

The MADS-box proteins function through the formation of homo- and heterodimers that bind conserved CArG boxes, whose consensus nucleotide sequence is $CC(A/T)_6GG$ [40]. After dimer formation, the MADS-box transcription factors further interact and constitute protein complexes (named the "floral quartets") that regulate floral organ-specific expression programs [55]. Based on this model, the complexes AP1/AP1/SEP/SEP, AP1/SEP/AP3/PI, AG/SEP/AP3/PI and AG/AG/SEP/SEP regulate the formation of the floral structures within whorls 1, 2, 3 and 4, respectively (Fig. **3A**) [55, 56].

The ABCDE model is generally well conserved [57-61]; however, functional and evolutionary analyses of the MADS-box genes in non-model species, such as orchids, have revealed interesting divergences in flower development mechanisms, which indicate that the functional role of the MADS-box genes is not always strictly related to structural orthology [62]. Tables 1 and 2 list some of the MADS-box genes characterized in orchids, Fig. (**4**) presents their evolutionary relationships, and the following sections describe their main functions.

The AP1/AGL9 Genes in Orchids

The functional classes A and E of the MADS-box genes are phylogenetically related and are generally referred as the *AP1/AGL9* group [46, 55]. Several duplication events resulted in the origin of this heterogeneous group of MADS-box genes [63, 64]. Within the *AP1/AGL9* group, the first division delineates the *AP1/SQUA*-like subfamily (from *APETALA1* of *Arabidopsis thaliana* and *SQUAMOSA* of *Antirrhinum majus*), which includes the class A genes. Based on the presence of specific conserved motifs in the C-terminal region, the AP1/SQUA-like proteins are further divided into two clades: the paleoAP1-like and the euAP1-like [65, 66]. The consensus sequence L/MPPWML is the paleoAP1 motif (or the FUL-like motif, from the *FRUITFULL* locus of *A. thaliana*) that characterizes the paleoAP1-like clade, whereas the euAP1-like clade presents two alternative consensus sequences, the euAP1 motif RRNaLaLT/NLa and the farnesylation motif CFAT/A, the latter of which originated from a frameshift mutation in the paleoAP1 motif [65, 66]. The roles of the paleoAP1 and farnesylation motifs are not clearly understood, and some AP1/SQUA-like proteins missing both motifs are not affected in their function [67, 68]. The *AP1/SQUA*-like genes have been found only in angiosperms, which supports the hypothesis that their emergence might be linked to the origin of the floral organs. The genes belonging to the eu*AP1*-like clade are present only in the higher eudicots, whereas those of the paleo*AP1* clade have been described in both monocot and dicot species [65, 66].

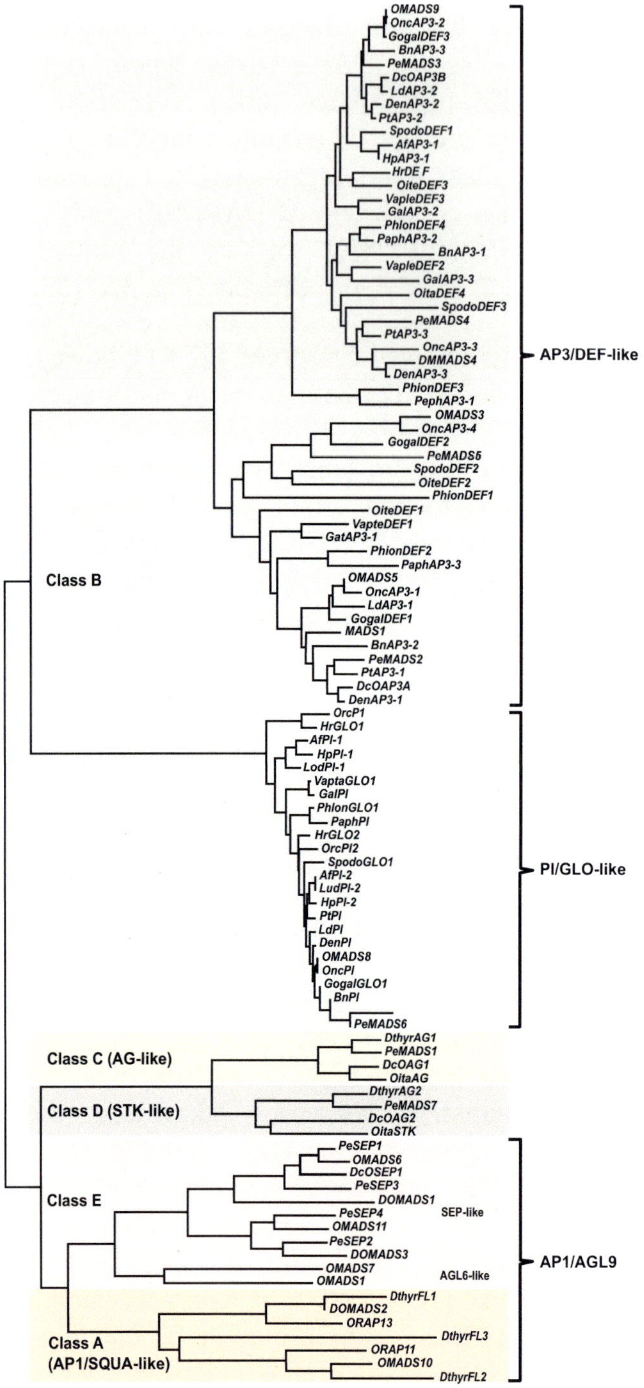

Figure 4: Neighbor-joining tree of the orchid MADS-box proteins showing the various functional classes.

The second main division of the *AP1/AGL9* genes is the *SEP*-like subfamily (from *SEPALLATA* of *A. thaliana*), which includes the E-function genes. Within this subfamily, two clades can be distinguished: *SEP3* and *SEP1/2/4* (previously referred to as *AGL9* and *AGL2/3/4*, respectively) [46, 63, 64, 69]. In addition, the *AP1/AGL9* group also includes the *AGL6* clade [70-73]. The determination of floral organs is not the only function of the *AP1/AGL9* genes; in fact, almost all of these genes are also involved in the initiation and development of the floral meristem [74], which suggests that the *AP1/AGL9* genes are at the top of the functional hierarchical scale of the MADS-box genes involved in flower development regulatory pathways [75-77].

The identification and functional characterization of a number of *AP1/AGL9* genes in orchids (Table **1**, Fig. **4**) have clarified many aspects of the molecular mechanisms of floral transition in this plant family.

Table 1. Some of the class A, C, D and E MADS-box genes identified in orchids.

Locus	Species	Clade	Class	Genbank Accession Number
DthyrFL1	*Dendrobium thyrsiflorum*	AP1/SQUA - like	A	AY927236
DthyrFL2	*Dendrobium thyrsiflorum*	AP1/SQUA - like	A	AY927237
DthyrFL3	*Dendrobium thyrsiflorum*	AP1/SQUA - like	A	AY927238
DOMADS2	*Dendrobium* Madame Thong-In	AP1/SQUA-like	A	AF198175
OMADS10	*Oncidium* Gower Ramsey	AP1/SQUA-like	A	HM140846
ORAP11	*Phalaenopsis amabilis*	AP1/SQUA-like	A	DQ104328
ORAP13	*Phalaenopsis amabilis*	AP1/SQUA-like	A	DQ104327
CeMADS1	*Cymbidium ensifolium*	AG-like	C	GU123626
CeMADS2	*Cymbidium ensifolium*	AG-like	C	GU123627
DcOAG1	*Dendrobium crumenatum*	AG-like	C	DQ119840
DthyrAG1	*Dendrobium thyrsiflorum*	AG-like	C	DQ017702
PeMADS1	*Phalaenopsis equestris*	AG-like	C	AF234617
DcOAG2	*Dendrobium crumenatum*	STK-like	D	DQ119841
DthyrAG2	*Dendrobium thyrsiflorum*	STK-like	D	DQ017703
PeMADS7	*Phalaenopsis equestris*	STK-like	D	JN983500
OMADS1	*Oncidium* Gower Ramsey	AGL6-like	E	HM140843
OMADS7	*Oncidium* Gower Ramsey	AGL6-like	E	HM140845

Table 1: contd….

DcOSEP1	*Dendrobium crumenatum*	SEP-like	E	DQ119842
DOMADS1	*Dendrobium* Madame Thong-In	SEP-like	E	AF198174
OMADS11	*Oncidium* Gower Ramsey	SEP-like	E	HM140847
DOMADS3	*Dendrobium* Madame Thong-In	SEP-like	E	AF198176
OMADS6	*Oncidium* Gower Ramsey	SEP-like	E	HM140844
PeSEP1	*Phalaenopsis equestris*	SEP-like	E	KF673857
PeSEP2	*Phalaenopsis equestris*	SEP-like	E	KF673858
PeSEP3	*Phalaenopsis equestris*	SEP-like	E	KF673859
PeSEP4	*Phalaenopsis equestris*	SEP-like	E	KF673860

In the orchid *Dendrobium* Madame Thong-In, *DOMADS1*, *DOMADS2* and *DOMADS3* are the MADS-box genes homologous to *SEP1*, *AP1/SQUA* and *SEP3*, respectively. During the floral transition of *Dendrobium*, the *DOMADS1-3* genes are activated in succession. Like the vast majority of the MADS-box genes that regulate the floral transition, *DOMADS1-3* continue to be expressed up to the stage of the mature flower [75] and also function in the stages of flower development subsequent to floral transition [78-81]. However, the expression profiles of the *DOMADS1-3* genes are quite divergent from those of the homologous genes in *Arabidopsis*, which demonstrates that there is no functional conservation in the MADS-box genes involved in floral transition. The gene *DOMADS1* is expressed in the inflorescence meristem, the floral primordium and all of the floral organs. The gene *DcOSEP1*, the ortholog of *DOMADS1* in *Dendrobium crumenatum*, has a similar expression profile in floral organs [82]. *DOMADS2* transcripts are detectable in the early apical meristem of the shoot and throughout the floral transition, whereas its late expression is limited to the column. The gene *DOMADS3* is transcribed before the differentiation of the flower primordium; later, its transcripts are restricted to the pedicel tissue.

DOMADS2 and *DOMADS3* are initially transcribed much earlier than their orthologs in *Arabidopsis*, in which the genes *AP1*, *AGL8* and *CAULIFLOWER* (*CAL*) [78, 83, 84] are expressed first, and the transcription of *SEP1*, *SEP4* and *SEP3* is activated later, in stage 2 of the flower primordium [85-87]. In addition to temporal expression pattern differences, the spatial expression profiles also vary. In fact, *DOMADS1* and *DOMADS2* are transcribed in both the inflorescence and the floral meristem, whereas in *Arabidopsis*, *AGL8* is expressed in the inflorescence and *AP1* and *CAL* in the floral meristem. These divergences in temporal and spatial gene expression patterns suggest the existence of different regulatory systems that have evolved specifically in various plant families to modulate the activity of the MADS-box genes involved in floral transition.

In addition to expression analysis studies, the characterization of gene regulatory regions is particularly useful for dissecting gene function. The promoter region of the *DOMADS1* gene has been examined, and many *cis*-acting regulatory elements affect its transcriptional activity in the reproductive organs and stem of *Dendrobium* [88]. Six CArG boxes, which are binding sites and key modulators of the expression of different MADS-box genes [89-92], are present within the promoter of the *DOMADS1* gene. These regulatory binding sites are commonly found in the promoters of MADS-box genes of distantly related species (*e.g.*, *Arabidopsis*), which indicates the conservation of the basic mechanisms that control the expression of the MADS-box genes through CArG-box sequence binding. Further regulatory sequences are present within the promoter of the *DOMADS1* gene; for example, there are five DNA-binding sites for *DOH1* [88], a class 1 *knox* gene that negatively modulates the transcriptional activity of *DOMADS1* during floral transition [93].

Other *AP1/AGL9* MADS-box genes have been isolated in the genus *Dendrobium*. The three genes of *D. thyrsiflorum DthyrFL1-3* belong to the paleo*AP1*-like group, which evolved from a single ancestor common to all monocots [68]. Both *DthyrFL1* and *DthyrFL2* have the paleoAP1-like motif, whereas this motif is absent in *DthyrFL3*. As in the evolutionary history of several MADS-box gene lineages [65, 94-96], the loss of the paleoAP1-like motif in *DthyrFL3* is likely due to a frameshift mutation. The transcript levels of the *DthyrFL1-3* genes are low in roots and leaves, moderate in ovules and high in inflorescences, in which the expression of *DthyrFL1* and *DthyrFL2* increases from small to large floral buds [68]. This expression profile suggests the involvement of these genes in different aspects of orchid inflorescence development.

The genes *ORAP11* and *ORAP13* in the orchid *Phalaenopsis amabilis* contain the paleo*AP1*-like motif, lack the farnesylation motif and belong to the *AP1/SQUA*-like subfamily [97]. Their early expression in the inflorescence and floral meristems of *P. amabilis* is comparable with that of the *FUL* gene in *A. thaliana* [98] and of the *OsMADS18* gene in rice [99], which supports the involvement of *ORAP11* and *ORAP13* in the establishment of meristem identity in this orchid species. During subsequent flower formation, both genes are transcribed in the primordia of all floral organs, which is in agreement with the expression patterns of the genes *OsMADS18* in *Oryza sativa* [99] and *LtMADS1* in *Lolium temulentum* [100] but not with those of the genes *AP1* and *FUL* in *Arabidopsis*

[83, 101]. Later, both *ORAP* genes are involved in the development of the tepals, lip, column and ovule, the last of which also involves *DthyrFL1-3* in *D. thyrsiflorum* [68] and *PFG* in petunia [102]. The transcription of *ORAP11* in the column of mature flowers of *P. amabilis* agrees with the expression profile of the *DOMADS2* gene in *Dendrobium* Madame Thong-In [75]. Transcripts of the *ORAP* genes are also present in the root and the procambial strand region, which is more in agreement with the expression patterns of the *FUL*, *PFG*, *OsMADS18* and *LtMADS1* genes than with those of *AP1*, *SQUA* or *PEAM4*, the *AP1* homolog in pea [100-106]. The transcriptional patterns of the *ORAP* genes led to the hypothesis that in orchids, the *AP1/SQUA*-like genes have maintained an ancestral function in the determination of meristem identity, but they have also evolved roles different from those of the "classic" class A genes.

The *AP1/AGL9* genes *PeSEP1-4* in the orchid *Phalaenopsis equestris* belong to the *SEP*-like clade and are transcribed in all floral organs; *PeSEP2* is also expressed in vegetative tissues. The PeSEP1-4 proteins can interact with other MADS-box factors of classes B, C, D and AGL6-like to establish floral organ identity. *PeSEP3* silencing causes the transformation of the tepals into leaf-like structures, whereas silencing of *PeSEP2* results in minor alterations of the floral phenotype. When expression of the class E genes *PeSEP2* and *PeSEP4* is silenced, the class B genes are downregulated, which reveals a link between the expression of the *PeSEP* genes and the class B genes [107].

OMADS1 is an *AP1/AGL9*-like gene in the orchid *Oncidium* Gower Ramsey and belongs to the *AGL6* clade [108]. Its expression is activated during the initial phase of apical meristem formation in *Oncidium*, functioning in a manner similar to that of other *AP1/AGL9* genes (*e.g.*, *AP1* and *SEP3*) to drive floral initiation [67, 109-111]. The OMADS1 protein can interact with the orchid class B MADS-box protein OMADS3 to form heterodimers that act in floral initiation [112]. In the mature flower of *Oncidium*, *OMADS1* is expressed in the lip and carpel, in contrast with the expression patterns of its orthologs *AGL6* in *Arabidopsis* and *ZAG3* in *Zea mays*, both of which are expressed in all four flower organs and ovules [70, 71]. OMADS1 can also form heterodimers with OMADS2, a class-D MADS-box protein whose transcripts are present in the stigmatic cavity and ovary of *Oncidium* [113].

In *Oncidium*, in addition to *OMADS1*, four *AP1/AGL9*-like genes have been characterized: *OMADS6*, *OMADS7*, *OMADS10* and *OMADS11* [114]. The *OMADS6* gene is an ortholog of *SEP3*, *OMADS7* is similar to the *AGL6*-like genes within the E-function genes, *OMADS10* is an orchid paleo*AP1* gene, and

OMADS11 is strictly related to the *SEP1/2* genes. *OMADS6*, *OMADS7* and *OMADS11* share overlapping expression profiles, whereas *OMADS10* exhibits an entirely different pattern. Unlike the *SEP3* gene and many of its orthologs, whose expression is restricted to the three inner floral whorls [44, 51, 64, 86, 115-118], *OMADS6* is transcribed in all four floral whorls and even weakly in stamens. This expression profile resembles that of *SEP1/2* [44, 87, 118] and *LMADS3*, a *Lilium SEP3* ortholog [119], and may be related to the presence of morphologically similar sepals and petals (tepals) in orchids and lilies. The genes *OMADS11* and *OMADS6*, neither of which are expressed in the stamens, have similar expression patterns. *OMADS7* and *OMADS6* have expression profiles comparable to those of *AGL6* in *A. thaliana* and *ZAG3* in maize [71]. In contrast, the expression pattern of *OMADS7* differs from that of *OMADS1*, whose transcripts are present in the lip and carpel [108]. The genes *OMADS6*, *OMADS7* and *OMADS11* have similar, although not identical, expression profiles, which suggests the existence of conserved transcriptional regulatory mechanisms.

Although almost all the genes of the *AP1/SQUA*-like subfamily are generally expressed in the early floral meristem and in floral organs, whereas their transcripts are not present in vegetative tissues [75, 83, 84, 103, 120, 121], some *AP1/SQUA*-like genes in monocots are also expressed in leaves [99, 122, 123]. The gene *OMADS10* in *Oncidium* is transcribed in the leaves, lips and carpels, which suggests a possible conserved function for specific lineages of *AP1/SQUA*-like genes in monocots.

The Class C and D Genes in Orchids

The ABCDE model of flower development attributes to the class C genes the regulation of the development of carpels and, together with the class B genes, of stamens. The class D genes are mainly implicated in ovule development. During the evolution of the angiosperms, an early duplication event gave rise to sister clades, the class C and D genes [95]. All class C and D MADS-box proteins contain two conserved sequences at their C-termini, the AG motifs I and II [95].

Although recent orchid next-generation sequencing (NGS) studies have identified many genes belonging to the C and D classes, the number of characterized C and D loci in Orchidaceae is smaller than those of the other classes (Table **1**, Fig. **4**).

In *Dendrobium crumenatum*, the MADS-box gene *DcOAG1* is a class C gene, and *DcOAG2* belongs to class D [82]. *DcOAG2* is a *SEEDSTICK* (*STK*) homolog specifically transcribed in the ovary. The gene *DcOAG1*, an ortholog of *AG* in *A.*

thaliana, has an additional amino acid stretch at the N-terminus, just upstream of the MADS domain, which is characteristic of the class C genes. The genomic organization of *DcOAG1* includes the presence of intron 8, which is commonly found in many class C genes and was probably lost in the class D clade after the divergence of Nymphaeales from other angiosperms [95, 124]. The expression of the gene *DcOAG1* is not restricted to reproductive tissues. In accordance with the expression of *AG* orthologs in some basal angiosperms, *DcOAG1* transcripts are present in all the floral organs [60]. The similar expression profiles of the *AG*-like genes in orchids and basal angiosperms suggest the possible independent evolution of the regulatory mechanisms involved in the expression of these class C genes.

The genes *DthyrAG1* and *DthyrAG2* in *Dendrobium thyrsiflorum* belong to classes C and D, respectively [125]. The conserved AG motifs I and II are present within the C-termini of both DthyrAG1 and DthyrAG2. In addition, DthyrAG2 contains the MD motif YET/AKA/DDXX, a typical extension of the AG motif found in the D lineage of monocots, which may play a role in the establishment of interactions with other regulatory proteins [95]. As also observed for *DcOAG1* in *D. crumenatum*, the genomic organization of the *DthyrAG1* gene in *D. thyrsiflorum* includes the presence of intron 8 before the stop codon. *DthyrAG1* and *DthyrAG2* transcripts are expressed during the development of the ovule and flower, in particular in the rostellum, stigma and stylar canal. Both in monocots and dicots, the class D genes are typically expressed in ovules [126, 127]; however, some exceptions have been observed in monocots. For example, the *LMADS2* gene in *Lilium longiflorum* is transcribed in the stylum [128], and the *ZmZAG2* gene in *Zea mays* is transcribed in the stigma [129]. The divergence of the class D MADS-box gene transcription profile between monocots and dicots and the specific presence of the MD motif in monocots may be related to the gain of novel additional functions for the class D genes within monocots. In agreement with the expression patterns of the class C and D genes in other species, both *DthyrAG1* and *DthyrAG2* are also transcribed during ovule development [44, 121, 126, 130, 131], although with some differences. *DthyrAG1* is only transcribed during the early stages of ovule development, whereas the expression of *DthyrAG2* remains active up to the end of this process, which suggests a role for *DthyrAG2* late in ovule development [125].

OMADS4 and *OMADS2* of *Oncidium* Gower Ramsey are class C and D MADS-box genes, respectively [113], both encoding proteins that contain AG motifs. In agreement with other class D proteins of monocots, OMADS2 also contains a typical MD motif. The *OMADS4* expression pattern is similar to that of other class C genes, being expressed only in stamens and carpels [129, 132, 133]. Additionally, the transcriptional profile of *OMADS2*, which is specifically expressed in carpels, is in agreement with that of other class D genes [134, 135].

In the orchid *Phalaenopsis equestris*, the genes *PeMADS1* and *PeMADS7* encode proteins that contain AG motifs in their C-termini, and PeMADS7 also contains the MD motif. *PeMADS1* and *PeMADS7* are class C and D MADS-box genes, respectively [136]. Both *PeMADS1* and *PeMADS7* are expressed during flower and ovule development, in accordance with the expression profiles of the class C and D genes of *Oncidium* and *Dendrobium* [113, 125]. In addition, both *PeMADS1* and *PeMADS7* are expressed in the placenta before ovule maturation, which in orchids is triggered by pollination. Thus, additional factors may be involved in the regulation of ovule initiation in orchids [136].

In *Orchis italica*, *OitaAG* and *OitaSTK* are class C and D genes, respectively. These genes are highly expressed in columns and ovaries, particularly after pollination. Surprisingly, the expression of *OitaAG* is weakly detectable in the inner tepals, and low levels of *OitaSTK* are expressed in the lip and root [137]. This expression pattern does not fully overlap with those described in other orchid species and may be a consequence of the different evolutionary histories of the class C and D genes in orchids. The intron 2 nucleotide sequence in *OitaAG* shares conserved *cis*-regulatory elements with the intron 2 of other *AG*-like genes. The presence of conserved binding sites for transcription factors that positively (*e.g.*, LEAFY and WUSCHEL) or negatively (*e.g.*, BELLRINGER) modulate the expression of *AG* homologs in dicots and monocots highlights the evolutionary conservation of these regulatory mechanisms [137].

In contrast with the orchid species that have only one MADS-box gene belonging to class C (*i.e.*, *Dendrobium thyrsiflorum*, *D. crumenatum*, *Oncidium*, *Phalaenopsis*, and *Orchis italica*), the orchid *Cymbidium ensifolium* has two class C genes: *CeMADS1* and *CeMADS2*. Both genes originated through a duplication event and encode regulatory proteins involved in gynostemium formation [138]. In the meristem tissue of *C. ensifolium*, *CeMADS1* and *CeMADS2* have redundant functions. However, in floral organs, these genes have temporally and spatially divergent expression profiles, which suggests that possible sub- and/or neo-functionalization might have occurred during the evolution of these two paralogs [138]. In agreement with the floral quartet model, the genes *CeMADS1* and *CeMADS2* exert their role during column development through the assembly of protein tetramers in which two CeMADS1 proteins or one CeMADS1 and one

CeMADS2 protein bind two class E proteins. The gene *CeMADS1* functions in stamen and carpel development, as revealed by the analysis of *multitepal*, a natural *Cymbidium* mutant that presents tepals in place of the column and expresses the *CeMADS2* gene but not *CeMADS1*. The expression of *CeMADS1* enhances the development of the column, and the expression of *CeMADS2* is necessary to conclude development in the correct manner, which indicates that *CeMADS2* is necessary but not sufficient to determine the formation of the column [138].

The Class B MADS-Box Genes in Orchids

In the ABCDE model, the functional class B MADS-box genes are involved in the development of the organs of the second and third floral whorls: petals and stamens. Within this group of genes, two main lineages originated through the duplication of an ancestral gene containing a paleoAP3 motif: the *AP3/DEF*-like lineage (from *APETALA3* in *A. thaliana* and *DEFICIENS* in *A. majus*) and the *PI/GLO*-like lineage (from *PISTILLATA* in *A. thaliana* and *GLOBOSA* in *A. majus*) [94, 139]. Within the *AP3/DEF*-like lineage, three groups can be distinguished: the paleoAP3, TM6 and euAP3 clades, the latter two of which arose from a second duplication event [94].

In orchids, the class B genes have been of particular interest, resulting in the characterization of a large number of genes from this functional class (Table **2**, Fig. **4**). Many class B MADS-box orchid genes exhibit an expression pattern that extends to the first whorl of the floral organs, which may be related to the presence of petaloid sepals in orchids (Fig. **3B**).

Table 2. Some of the class B MADS-box genes identified in orchids.

Locus	Species	Clade	Class	Genbank Accession Number
AfAP3-1	*Anoectochilus formosanus*	AP3/DEF-like	B	EU444018
BnAP3-1	*Brassavola nodosa*	AP3/DEF-like	B	EU444021
BnAP3-2	*Brassavola nodosa*	AP3/DEF-like	B	EU444022
BnAP3-3	*Brassavola nodosa*	AP3/DEF-like	B	EU444023
MADS1	*Cymbidium* hybrid cultivar.	AP3/DEF-like	B	DQ683575
DcOAP3A	*Dendrobium crumenatum*	AP3/DEF-like	B	DQ119838
DcOAP3B	*Dendrobium crumenatum*	AP3/DEF-like	B	DQ119839
DMMADS4	*Dendrobium moniliforme*	AP3/DEF-like	B	GU132995
DenAP3-1	*Dendrobium* Spring Jewel	AP3/DEF-like	B	EU444025
DenAP3-2	*Dendrobium* Spring Jewel	AP3/DEF-like	B	EU444026
DenAP3-3	*Dendrobium* Spring Jewel	AP3/DEF-like	B	EU444027

Table 2: contd…

GogalDEF1	*Gongora galeata*	AP3/DEF-like	B	FJ804097
GogalDEF2	*Gongora galeata*	AP3/DEF-like	B	FJ804098
GogalDEF3	*Gongora galeata*	AP3/DEF-like	B	FJ804099
HpAP3-1	*Habenaria petelotii*	AP3/DEF-like	B	EU444033
HrDEF	*Habenaria radiata*	AP3/DEF-like	B	AB232663
LdAP3-1	*Liparis distans*	AP3/DEF-like	B	EU444036
LdAP3-2	*Liparis distans*	AP3/DEF-like	B	EU444037
OMADS3	*Oncidium* Gower Ramsey	AP3/DEF-like	B	AY196350
OMADS5	*Oncidium* Gower Ramsey	AP3/DEF-like	B	HM140840
OMADS9	*Oncidium* Gower Ramsey	AP3/DEF-like	B	HM140841
OncAP3-1	*Oncidium* Gower Ramsey	AP3/DEF-like	B	EU444041
OncAP3-2	*Oncidium* Gower Ramsey	AP3/DEF-like	B	EU444042
OncAP3-3	*Oncidium* Gower Ramsey	AP3/DEF-like	B	EU444043
OncAP3-4	*Oncidium* Gower Ramsey	AP3/DEF-like	B	EU444044
OitaDEF1	*Orchis italica*	AP3/DEF-like	B	AB857726
OitaDEF2	*Orchis italica*	AP3/DEF-like	B	AB857727
OitaDEF3	*Orchis italica*	AP3/DEF-like	B	AB857728
OitaDEF4	*Orchis italica*	AP3/DEF-like	B	AB857729
PtAP3-1	*Phaius tankervilleae*	AP3/DEF-like	B	EU444050
PtAP3-2	*Phaius tankervilleae*	AP3/DEF-like	B	EU444051
PtAP3-3	*Phaius tankervilleae*	AP3/DEF-like	B	EU444052
PeMADS2	*Phalaenopsis equestris*	AP3/DEF-like	B	AY378149
PeMADS3	*Phalaenopsis equestris*	AP3/DEF-like	B	AY378150
PeMADS4	*Phalaenopsis equestris*	AP3/DEF-like	B	AY378147
PeMADS5	*Phalaenopsis equestris*	AP3/DEF-like	B	AY378148
PaphAP3-1	*Paphiopedilum* Macabre	AP3/DEF-like	B	EU444046
PaphAP3-2	*Paphiopedilum* Macabre	AP3/DEF-like	B	EU444047
PaphAP3-3	*Paphiopedilum* Macabre	AP3/DEF-like	B	EU444048
PhlonDEF2	*Phragmipedium longiflorum*	AP3/DEF-like	B	FJ804106
PhlonDEF1	*Phragmipedium longiflorum*	AP3/DEF-like	B	FJ804105
PhlonDEF3	*Phragmipedium longiflorum*	AP3/DEF-like	B	FJ804107
PhlonDEF4	*Phragmipedium longiflorum*	AP3/DEF-like	B	FJ804108
SpodoDEF2	*Spiranthes odorata*	AP3/DEF-like	B	FJ804111
SpodoDEF1	*Spiranthes odorata*	AP3/DEF-like	B	FJ804110
SpodoDEF3	*Spiranthes odorata*	AP3/DEF-like	B	FJ804112
GalAP3-1	*Vanilla pilifera*	AP3/DEF-like	B	EU444029

Table 2: contd…

GalAP3-2	*Vanilla pilifera*	AP3/DEF-like	B	EU444030
GalAP3-3	*Vanilla pilifera*	AP3/DEF-like	B	EU444031
VaplaDEF1	*Vanilla planifolia*	AP3/DEF-like	B	FJ804115
VaplaDEF3	*Vanilla planifolia*	AP3/DEF-like	B	FJ804117
VaplaDEF2	*Vanilla planifolia*	AP3/DEF-like	B	FJ804116
AfPI-1	*Anoectochilus formosanus*	PI/GLO-like	B	EU444019
AfPI-2	*Anoectochilus formosanus*	PI/GLO-like	B	EU444020
BnPI	*Brassavola nodosa*	PI/GLO-like	B	EU444024
DenPI	*Dendrobium* Spring Jewel	PI/GLO-like	B	EU444028
EpalPI	*Epipactis palustris*	PI/GLO-like	B	DQ005588
GogalGLO1	*Gongora galeata*	PI/GLO-like	B	FJ804100
HpPI-1	*Habenaria petelotii*	PI/GLO-like	B	EU444034
HpPI-2	*Habenaria petelotii*	PI/GLO-like	B	EU444035
HrGLO1	*Habenaria radiata*	PI/GLO-like	B	AB232665
HrGLO2	*Habenaria radiata*	PI/GLO-like	B	AB232664
LdPI	*Liparis distans*	PI/GLO-like	B	EU444038
LudPI-1	*Ludisia discolor*	PI/GLO-like	B	EU444039
LudPI-2	*Ludisia discolor*	PI/GLO-like	B	EU444040
OMADS8	*Oncidium* Gower Ramsey	PI/GLO-like	B	HM140842
OncPI	*Oncidium* Gower Ramsey	PI/GLO-like	B	EU444045
OrcPI	*Orchis italica*	PI/GLO-like	B	AB094985
OrcPI2	*Orchis italica*	PI/GLO-like	B	AB537504
PtPI	*Phaius tankervilleae*	PI/GLO-like	B	EU444053
PeMADS6	*Phalaenopsis equestris*	PI/GLO-like	B	AY678299
PaphPI	*Paphiopedilum* Macabre	PI/GLO-like	B	EU444049
PhlonGLO1	*Phragmipedium longiflorum*	PI/GLO-like	B	FJ804109
SpodoGLO1	*Spiranthes odorata*	PI/GLO-like	B	FJ804114
GalPI	*Vanilla pilifera*	PI/GLO-like	B	EU444032
VaplaGLO1	*Vanilla planifolia*	PI/GLO-like	B	FJ804118

The genes *PeMADS2-5* in *Phalaenopsis equestris* belong to the *AP3/DEF*-like lineage. Their expression is maintained throughout early- to late-stage inflorescence development, with an organ-specific pattern that reveals the absence of functional redundancy [140]. Indeed, *PeMADS2* transcription is very high in the outer and inner tepals and lower in the column, whereas *PeMADS3* is strongly expressed in the inner tepals and lips and more weakly expressed in the column. The *PeMADS4* transcripts are restricted to the lips and the column. Finally, the *PeMADS5* expression pattern expands from the tepals (outer, inner and lip) to the column. The functional role of the *PeMADS2-5* genes has been inferred through an analysis of their expression profiles in a natural peloric mutant of *Phalaenopsis*. *PeMADS2* is mainly involved in the development of the outer tepals, *PeMADS4* in the lip and column formation, and *PeMADS5* in the inner tepals and stamen development [140]. The sequence diversity of the *PeMADS2-5* promoter regions has contributed to their sub-functionalization [141]. In addition, histone acetylation (H3K9K14ac) in the region surrounding the *PeMADS4* translation start site is higher in the lip than in the other floral tissues. Thus, the upstream *cis*-regulatory regions and the modulation of histone acetylation may determine the distinct expression profiles of the *AP3/DEF*-like genes during *Phalaenopsis* floral development [141].

Unlike the *AP3/DEF*-like lineage of *P. equestris*, which includes four genes, the *PI/GLO*-like lineage contains only the *PeMADS6* gene [142]. This gene is expressed in the inflorescence meristem and floral primordium, which suggests its involvement in early floral development. Furthermore, *PeMADS6* is also transcribed in all the organs of the inflorescence (outer and inner tepals, lip, column and ovary), which reveals its role in the development of the floral organs. In addition, *PeMADS6* expression in the flower organs persists until senescence, and this pattern might be related to the involvement of this gene in flower longevity [142]. The AP3/DEF-like proteins PeMADS2-5 are able to bind the PeMADS6 protein to regulate the formation of specific organs [143]. Furthermore, the self-binding capability of PeMADS4 and PeMADS6 determines the formation of PeMADS4 homodimers and PeMADS6 homodimers/homomultimers that recognize the CArG boxes. These conserved MADS-box protein DNA-binding motifs are also recognized by the heterodimers PeMADS2–PeMADS6, PeMADS4–PeMADS6 and PeMADS5–PeMADS6, which indicates that the orchid *AP3/DEF*-like and *PI/GLO*-like proteins can be combined in different manners to exert complex regulatory functions [143].

DcOAP3A and *DcOAP3B* in *Dendrobium crumenatum* are *AP3/DEF*-like genes that belong to the paleo*AP3* lineage, and *DcOPI* is a *PI/GLO*-like gene [82]. *DcOPI* and *DcOAP3A* are transcribed in all the floral whorls, whereas *DcOAP3B* transcripts are present in the inner tepals, lip, column and pollinia. In addition, all three genes are expressed in the ovary [82].

The three class B MADS-box genes described in the orchid *Habenaria radiata* include two *PI/GLO*-like genes (*HrGLO1* and *HrGLO2*) and one *AP3/DEF*-like gene (*HrDEF*) [144]. *HrGLO1* and *HrGLO2* are expressed in the outer and inner tepals and in the column, whereas the expression of *HrDEF* is detectable only in the inner tepals and the column [144].

The gene *OMADS8* in *Oncidium* Gower Ramsey belongs to the *PI/GLO*-like lineage, whereas *OMADS3*, *OMADS5* and *OMADS9* are *AP3/DEF*-like genes [112, 145]. *OMADS8* transcripts are present in all the flower organs and in the leaf tissue. Additionally, *OMADS3* is expressed in all four floral whorls and in leaves, similarly to several *AP3/DEF*-like genes in the TM6 clade [146]. *OMADS5* transcription is restricted to the tepals (outer and inner), in agreement with the expression pattern of *PeMADS2* in *Phalaenopsis* [140]. Finally, *OMADS9* is expressed in the inner tepals and lip, in accordance with the transcriptional profiles of the genes *DcOAP3B*, *PeMADS3* and *HrDEF* [82, 140, 144]. The expression pattern of *OMADS3*, transcribed in all floral organs and leaves, is different from that of *OMADS5* and *OMADS9,* whose transcripts are absent from stamens and leaves [112]. These divergences suggest the functional diversification of the three *AP3/DEF*-like genes in *Oncidium* [145]. The proteins OMADS3, OMADS5 and OMADS9 can form homodimers; in addition, OMADS3 can interact with OMADS8, OMADS5 and OMADS9 to form heterodimers, and OMADS5 can form heterodimers with OMADS9 [145].

The *OrcPI* gene in *Orchis italica* belongs to the class B *PI/GLO*-like lineage [147]. Its transcripts are present in all floral whorls, from early- to late-stage inflorescence formation and senescence, which confirms the involvement of the *PI/GLO*-like genes in orchid flower longevity, as also observed in *Phalaenopsis* [148]. Comparative evolutionary analyses have examined the selective constraints acting on the MADS-box loci to identify eventual signals of neutral, purifying and adaptive selection. The *OrcPI* locus is subject to heterogeneous selective pressures, with various evolutionary constraints acting on its coding and non-coding regions [147, 148]. Furthermore, the 5' regulatory sequence of *OrcPI* shares conserved regulatory motifs with the corresponding regions of homologous genes in *Oryza sativa*, *Lilium regale* and *Arabidopsis thaliana*, which confirms the strong conservation of the regulatory signals involved in the development of floral tissues [149]. In several species of the Orchidoideae subfamily, two *PI/GLO*-like genes have been identified [150-152]. For example, the *OrcPI2* gene in *O. italica* is a paralog of *OrcPI*. These class B *PI/GLO*-like genes have distinct expression levels and are subject to different selective pressures. In particular, the evolutionary pattern of their synonymous sites reveals that possible sub-functionalization events may have occurred after the divergence of these two paralogs [151, 152]. In *O. italica*, four genes of the *AP3/DEF*-like lineage have also been identified (*OitaDEF1-4*) [27]. *OitaDEF1* is transcribed mainly in the outer tepals, whereas the gene *OitaDEF2* is highly expressed in the outer tepals and at lower levels in the inner tepals and lip. The expression levels of *OitaDEF3* and *OitaDEF4* are higher in the inner tepals and lip than in the other organs. Interestingly, the *OitaDEF2* transcript is regulated by the microRNA miR5179 through cleavage at a specific nucleotide position [27].

The identification of a large number of class B genes from several species belonging to different Orchidaceae subfamilies revealed the existence of four distinct clades (from 1 to 4) within the *AP3/DEF*-like lineage, whereas the *PI/GLO*-like lineage includes paralogous genes only in the Orchidoideae subfamily, in which they originated as a consequence of a recent duplication event [144, 150, 151, 153]. The genes of the *AP3/DEF*-like clade 2 lineage are subject to relaxed evolutionary constraints as compared with other functional class B MADS-box genes in orchids. The sub- and neo-functionalization events that occurred after gene duplication within the orchid *AP3/DEF*-like lineage have played a pivotal role in the evolution of floral morphological traits, determining the extreme diversification of the organs of the perianth in this plant family [153].

The Orchid Code

In orchids and other monocot species, the first and second floral whorls are morphologically similar organs (the tepals), whereas the outer whorls of the model species *Arabidopsis* and other eudicots are phenotypically different organs (the sepals in the first whorl and the petals in the second whorl). The existence of tepaloid sepals and petals is attributed to the extension of the transcriptional domain of the class B MADS-box genes into whorl 1, as described in the modified ABCDE model of flower development [154, 155] (Fig. **3B**). However, this modified model cannot convincingly explain the morphological diversity of the orchid tepals, which are distinct in the outer, lateral inner and median inner (lip), the last of which exhibits highly specialized features.

Over the last decade, knowledge concerning the molecular mechanisms controlling flower development in orchids has greatly increased. Nevertheless, questions still remain concerning the origin and evolution of the morphological diversification of the orchid floral organs. An attempt to harmonize the morphological, developmental and molecular data obtained from various orchid species resulted in the proposed "orchid code", an evolutionary model that explains the evolutionary origin and peculiarities of the orchid flower [6, 156, 157].

The developmental-genetic code proposed by the orchid code theory designates the class B *AP3/DEF*-like genes as the major players in the establishment of tepal and lip identity, whereas the functions of the class B *PI/GLO*-like genes and those of the class A, C, D and E genes are in substantial agreement with the modified ABCDE model.

In contrast with *Arabidopsis* and other eudicot species, in which the binding of one *AP3/DEF*-like and one *PI/GLO*-like gene product is essential to determine petal identity, the orchid code theory proposes that the identity of tepals and lips is established through the formation of protein complexes among the four different orchid AP3/DEF-like proteins and one PI/GLO-like protein. The orchid class B *AP3/DEF*-like lineage includes clade 1 (*PeMADS2*-like), which is sister to clade 2 (*OMADS3*-like), and cladc 3 (*PeMADS3*-like), which is sister to clade 4 (*PeMADS4*-like). The genes belonging to each *AP3/DEF*-like clade display a specific expression profile [156, 157].

The orchid code theory proposes that the interaction of clade 1 and clade 2 proteins in whorl 1 drives the formation of the outer tepals. In whorl 2, the development of the two lateral inner tepals is determined by the presence of high levels of clade 1 and 2 proteins and low levels of clade 3 and 4 proteins, whereas the formation of the modified inner median tepal (lip) is specified by high levels of clade 3 and 4 proteins and low levels of clade 1 and 3 proteins (Fig. **5**). In brief, the proteins of clade 3 distinguish the inner from the outer tepals, whereas the proteins of clade 4 distinguish the lateral inner tepals from the lip [6, 156].

This model also illustrates the evolution of the bilaterally symmetric orchid flower starting from a flower with radial symmetry that possessed six very similar tepals, in which the single common ancestor of the current orchid *AP3/DEF*-like genes was uniformly transcribed. As in several gene families, subsequent duplications, followed by the divergence of *cis*-regulatory motifs, were crucial events in the evolution of the four *AP3/DEF*-like orchid clades. The first duplication gave rise to the ancestor of the clade 1 and 2 genes and the ancestor of the clade 3 and 4 genes. Subsequent specialization in the expression patterns of the ancestor of the clade 3 and 4 genes, which are not transcribed in the outer tepals, might have led to the evolution of an intermediate flower organization with distinguishable outer and inner tepals (Fig. **5**). A second duplication event triggered the diversification of the clade 3 and 4 genes, and the modularization of their transcriptional patterns resulted in the evolution of the lip [156, 157].

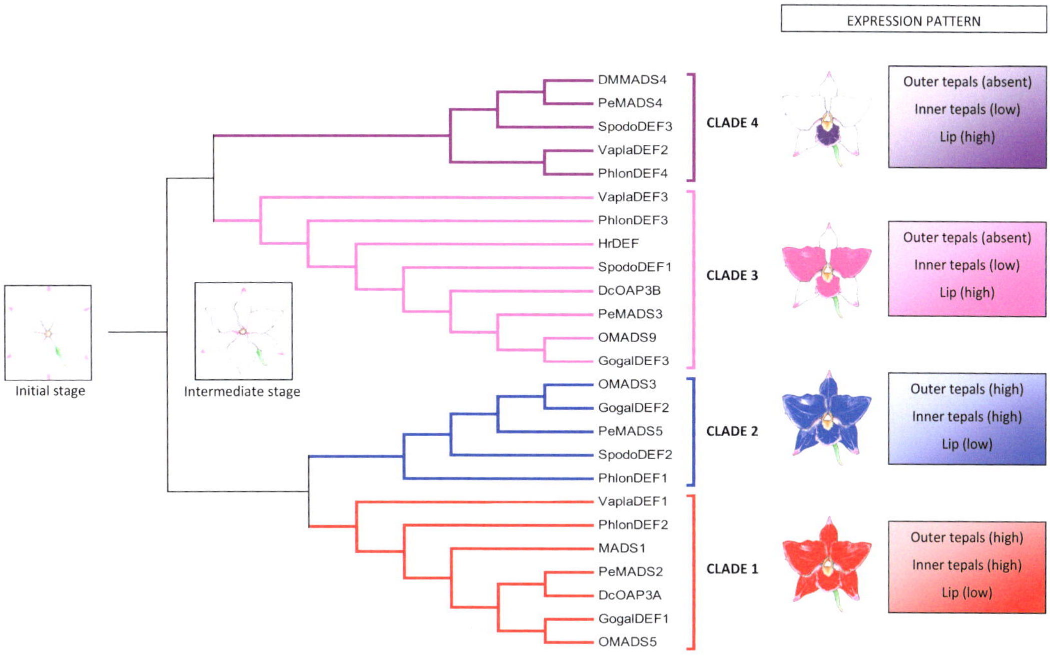

Figure 5: The orchid code model, modified from Mondragon-Palomino and Theissen [156, 157]. The colors denote the various clades of the *AP3/DEF*-like genes and their expression profiles in the orchid floral organs. A model for the possible evolutionary stages of the orchid perianth is also presented.

CONCLUDING REMARKS

Recent advances in molecular evolutionary studies have deeply increased our understanding of the "abominable mystery" of the orchid flower. Without doubt, future studies of the orchid MADS-box genes involved in flower development will benefit greatly from NGS analyses. Constantly accumulating data concerning the structure, function and evolution of the genes involved in orchid flower development have uncovered many aspects of orchid development. Although the evolutionary role of the recently discovered paralogs of the *PI/GLO*-like genes is still not understood, the elegant orchid code theory successfully explains the complex role of the class B MADS-box genes in the evolution of the orchid perianth. A more comprehensive analysis of the orchid MADS-box genes belonging to the other functional classes (A, C, D and E) will lead to a more complete model that may also describe the evolutionary origin of the orchid's reproductive structures.

CONFLICT OF INTEREST

The author confirms that he has no conflict of interest to declare for this publication.

ACKNOWLEDGEMENTS

The authors are deeply grateful to Dr. Marco Salvemini for the elaboration of the orchid flower schemes in (Figs. **1** and **5**).

DISCLOSURE

This is an updated chapter of the earlier review article [by S. Accto & L. Gaudio, "The MADS and the beauty: Genes involved in the development of orchid flowers", *Curr. Genomics*, 2011, 12(5): 342-356], with additional text and references.

REFERENCES

[1] Cozzolino S, Widmer A. Orchid diversity: an evolutionary consequence of deception? Trends Ecol Evol 2005; 20(9): 487-94.

[2] Tremblay RL, Ackerman JD, Zimmerman JK, Calvo RN. Variation in sexual reproduction in orchids and its evolutionary consequences: a spasmodic journey to diversification. Biol J Linn Soc 2005; 84(1): 1-54.

[3] Yu H, Goh CJ. Molecular genetics of reproductive biology in orchids. Plant Physiol 2001; 127(4): 1390-3.

[4] Tsai WC, Chen HH. The orchid MADS-box genes controlling floral morphogenesis. ScientificWorldJournal 2006; 6: 1933-44.

[5] Rudall PJ, Bateman RM. Roles of synorganisation, zygomorphy and heterotopy in floral evolution: the gynostemium and labellum of orchids and other lilioid monocots. Biol Rev Camb Philos Soc 2002; 77(3): 403-41.

[6] Mondragon-Palomino M, Theissen G. Why are orchid flowers so diverse? Reduction of evolutionary constraints by paralogues of class B floral homeotic genes. Ann Bot (Lond) 2009; 104(3): 583-94.

[7] Cameron KM. Molecular phylogenetics of Orchidaceae: the first decade of DNA sequencing. In Cameron KM, Arditti J, Kull T, Eds. Orchid biology: reviews and perspectives. 9th ed. New York: New York Botanical Garden Press 2007; pp. 163–200.

[8] Cameron KM, Chase MW, Whitten WM, *et al.* A phylogenetic analysis of the Orchidaceae: evidence from rbcL nucleotide sequences. Am J Bot 1999; 86(2): 208-24.

[9] Gorniak M, Paun O, Chase MW. Phylogenetic relationships within Orchidaceae based on a low-copy nuclear coding gene, *Xdh*: Congruence with organellar and nuclear ribosomal DNA results. Mol Phylogenet Evol 2010; 56(2): 784-95.

[10] Wikstrom N, Savolainen V, Chase MW. Evolution of the angiosperms: calibrating the family tree. P Roy Soc B-Biol Sci 2001; 268(1482): 2211-20.

[11] Bremer K. Early Cretaceous lineages of monocot flowering plants. P Natl Acad Sci USA 2000; 97(9): 4707-11.

[12] Ramirez SR, Gravendeel B, Singer RB, Marshall CR, Pierce NE. Dating the origin of the Orchidaceae from a fossil orchid with its pollinator. Nature 2007; 448(7157): 1042-5.

[13] Janssen T, Bremer K. The age of major monocot groups inferred from 800+*rbcL* sequences. Bot J Linn Soc 2004; 146(4): 385-98.

[14] Gustafsson AL, Verola CF, Antonelli A. Reassessing the temporal evolution of orchids with new fossils and a Bayesian relaxed clock, with implications for the diversification of the rare South American genus *Hoffmannseggella* (Orchidaceae: Epidendroideae). BMC Evol Biol 2010; 10: 177.

[15] Conran JG, Bannister JM, Lee DE. Earliest orchid macrofossils: early Miocene *Dendrobium* and *Earina* (Orchidaceae: Epidendroideae) from New Zealand. Am J Bot 2009; 96(2): 466-74.

[16] Peakall R. Speciation in the Orchidaceae: confronting the challenges. Mol Ecol 2007; 16(14): 2834-7.

[17] Cai J, Liu X, Vanneste K, *et al.* The genome sequence of the orchid *Phalaenopsis equestris*. Nat Genet 2015; 47(1): 65-72.

[18] Su CL, Chao YT, Alex Chang YC, *et al.* *De novo* assembly of expressed transcripts and global analysis of the *Phalaenopsis aphrodite* transcriptome. Plant Cell Physiol 2011; 52(9): 1501-14.

[19] Chang YY, Chu YW, Chen CW, *et al.* Characterization of *Oncidium* 'Gower Ramsey' transcriptomes using 454 GS-FLX pyrosequencing and their application to the identification of genes associated with flowering time. Plant Cell Physiol 2011; 52(9): 1532-45.

[20] An FM, Chan MT. Transcriptome-wide characterization of miRNA-directed and non-miRNAdirected endonucleolytic cleavage using degradome analysis under low ambient temperature in *Phalaenopsis aphrodite* subsp. *formosana*. Plant Cell Physiol 2012; 53(10): 1737-50.

[21] Chou ML, Shih MC, Chan MT, *et al.* Global transcriptome analysis and identification of a *CONSTANS*-like gene family in the orchid *Erycina pusilla*. Planta 2013; 237(6): 1425-41.

[22] Zhang J, Wu K, Zeng S, *et al.* Transcriptome analysis of *Cymbidium sinense* and its application to the identification of genes associated with floral development. BMC Genomics 2013; 14: 279.

[23] Sedeek KE, Qi W, Schauer MA, *et al.* Transcriptome and proteome data reveal candidate genes for pollinator attraction in sexually deceptive orchids. PLoS One 2013; 8(5): e64621.

[24] Chao YT, Su CL, Jean WH, *et al.* Identification and characterization of the microRNA transcriptome of a moth orchid *Phalaenopsis aphrodite*. Plant Mol Biol 2014; 84(4-5): 529-48.

[25] Li X, Luo J, Yan T, *et al.* Deep sequencing-based analysis of the *Cymbidium ensifolium* floral transcriptome. PLoS One 2013; 8(12): e85480.

[26] De Paolo S, Salvemini M, Gaudio L, Aceto S. *De novo* transcriptome assembly from inflorescence of *Orchis italica*: analysis of coding and non-coding transcripts. PLoS One 2014; 9(7): e102155.

[27] Aceto S, Sica M, De Paolo S, *et al.* The analysis of the inflorescence miRNome of the orchid *Orchis italica* reveals a *DEF*-like MADS-box gene as a new miRNA target. PLoS One 2014; 9(5): e97839.

[28] Zhao X, Zhang J, Chen C, *et al.* Deep sequencing-based comparative transcriptional profiles of *Cymbidium hybridum* roots in response to mycorrhizal and non-mycorrhizal beneficial fungi. BMC Genomics 2014; 15: 747.

[29] Rao X, Krom N, Tang Y, *et al.* A deep transcriptomic analysis of pod development in the vanilla orchid (*Vanilla planifolia*). BMC Genomics 2014; 15: 964.

[30] Lin CS, Hsu CT, Liao C, *et al.* Transcriptome-wide analysis of the MADS-box gene family in the orchid *Erycina pusilla*. Plant Biotechnol J 2015; 1–15. doi: 10.1111/pbi.12383.

[31] Huang JZ, Lin CP, Cheng TC, *et al.* A *de novo* floral transcriptome reveals clues into *Phalaenopsis* orchid flower development. PLoS One 2015; 10(5): e0123474.

[32] Fu CH, Chen YW, Hsiao YY, *et al.* OrchidBase: A collection of sequences of the transcriptome derived from orchids. Plant Cell Physiol 2011; 52(2): 238-43.

[33] Tsai WC, Fu CH, Hsiao YY, *et al.* OrchidBase 2.0: comprehensive collection of Orchidaceae floral transcriptomes. Plant Cell Physiol 2013; 54(2): e7.

[34] Su CL, Chao YT, Yen SH, *et al.* Orchidstra: an integrated orchid functional genomics database. Plant Cell Physiol 2013; 54(2): e11.

[35] Schwarz-Sommer Z, Huijser P, Nacken W, Saedler H, Sommer H. Genetic control of flower development by homeotic genes in *Antirrhinum majus*. Science 1990; 250(4983): 931-6.

[36] Gramzow L, Ritz MS, Theissen G. On the origin of MADS-domain transcription factors. Trends Genet 2010; 26(4): 149-53.

[37] Alvarez-Buylla ER, Pelaz S, Liljegren SJ, *et al.* An ancestral MADS-box gene duplication occurred before the divergence of plants and animals. Proc Natl Acad Sci U S A 2000; 97(10): 5328-33.

[38] Parenicova L, de Folter S, Kieffer M, *et al.* Molecular and phylogenetic analyses of the complete MADS-box transcription factor family in *Arabidopsis*: new openings to the MADS world. Plant Cell 2003; 15(7): 1538-51.

[39] Masiero S, Colombo L, Grini PE, Schnittger A, Kater MM. The emerging importance of type I MADS box transcription factors for plant reproduction. Plant Cell 2011; 23(3): 865-72.

[40] Riechmann JL, Meyerowitz EM. MADS domain proteins in plant development. Biol Chem 1997; 378(10): 1079-101.

[41] Kaufmann K, Melzer R, Theissen G. MIKC-type MADS-domain proteins: structural modularity, protein interactions and network evolution in land plants. Gene 2005; 347(2): 183-98.

[42] Henschel K, Kofuji R, Hasebe M, *et al.* Two ancient classes of MIKC-type MADS-box genes are present in the moss *Physcomitrella patens*. Mol Biol Evol 2002; 19(6): 801-14.

[43] Zobell O, Faigl W, Saedler H, Munster T. MIKC* MADS-box proteins: conserved regulators of the gametophytic generation of land plants. Mol Biol Evol 2010; 27(5): 1201-11.
[44] Rounsley SD, Ditta GS, Yanofsky MF. Diverse roles for MADS box genes in *Arabidopsis* development. Plant Cell 1995; 7(8): 1259-69.
[45] Alvarez-Buylla ER, Liljegren SJ, Pelaz S, *et al.* MADS-box gene evolution beyond flowers: expression in pollen, endosperm, guard cells, roots and trichomes. Plant J 2000; 24(4): 457-66.
[46] Theissen G. Development of floral organ identity: stories from the MADS house. Curr Opin Plant Biol 2001; 4(1): 75-85.
[47] Becker A, Theissen G. The major clades of MADS-box genes and their role in the development and evolution of flowering plants. Mol Phylogenet Evol 2003; 29(3): 464-89.
[48] Theissen G, Melzer R. Molecular mechanisms underlying origin and diversification of the angiosperm flower. Ann Bot 2007; 100(3): 603-19.
[49] Soltis DE, Soltis PS, Albert VA, *et al.* Missing links: the genetic architecture of flower and floral diversification. Trends Plant Sci 2002; 7(1): 22-31; dicussion 31-4. Erratum in Trends Plant Sci 2002; 7(2): 95.
[50] Zhao L, Kim Y, Dinh TT, Chen X. miR172 regulates stem cell fate and defines the inner boundary of *APETALA3* and *PISTILLATA* expression domain in *Arabidopsis* floral meristems. Plant J 2007; 51(5): 840-9.
[51] Shitsukawa N, Tahira C, Kassai K, *et al.* Genetic and epigenetic alteration among three homoeologous genes of a class E MADS box gene in hexaploid wheat. Plant Cell 2007; 19(6): 1723-37.
[52] Gramzow L, Theissen G. A hitchhiker's guide to the MADS world of plants. Genome Biol 2010; 11(6): 214.
[53] Coen ES, Meyerowitz EM. The war of the whorls - Genetic interactions controlling flower development. Nature 1991; 353(6339): 31-37.
[54] Krizek BA, Fletcher JC. Molecular mechanisms of flower development: an armchair guide. Nat Rev Genet 2005; 6(9): 688-98.
[55] Theissen G, Saedler H. Plant biology. Floral quartets. Nature 2001; 409(6819): 469-71.
[56] Liu ZC, Mara C. Regulatory mechanisms for floral homeotic gene expression. Semin Cell Dev Biol 2010; 21(1): 80-86.
[57] Ambrose BA, Lerner DR, Ciceri P, *et al.* Molecular and genetic analyses of the *silky1* gene reveal conservation in floral organ specification between eudicots and monocots. Mol Cell 2000; 5(3): 569-79.
[58] Whipple CJ, Ciceri P, Padilla CM, *et al.* Conservation of B-class floral homeotic gene function between maize and *Arabidopsis*. Development 2004; 131(24): 6083-91.
[59] Whipple CJ, Zanis MJ, Kellogg EA, Schmidt RJ. Conservation of B class gene expression in the second whorl of a basal grass and outgroups links the origin of lodicules and petals. Proc Natl Acad Sci U S A 2007; 104(3): 1081-6.
[60] Kim S, Koh J, Yoo MJ, *et al.* Expression of floral MADS-box genes in basal angiosperms: implications for the evolution of floral regulators. Plant J 2005; 43(5): 724-44.
[61] Ferrario S, Immink RG, Angenent GC. Conservation and diversity in flower land. Current Opinion in Plant Biology 2004; 7(1): 84-91.
[62] Irish VF, Litt A. Flower development and evolution: gene duplication, diversification and redeployment. Curr Opin Genet Dev 2005; 15(4): 454-60.
[63] Malcomber ST, Kellogg EA. *SEPALLATA* gene diversification: brave new whorls. Trends Plant Sci 2005; 10(9): 427-35.
[64] Zahn LM, Kong H, Leebens-Mack JH, *et al.* The evolution of the *SEPALLATA* subfamily of MADS-box genes: a preangiosperm origin with multiple duplications throughout angiosperm history. Genetics 2005; 169(4): 2209-23.
[65] Litt A, Irish VF. Duplication and diversification in the *APETALA1/FRUITFULL* floral homeotic gene lineage: implications for the evolution of floral development. Genetics 2003; 165(2): 821-33.
[66] Vandenbussche M, Theissen G, Van de Peer Y, Gerats T. Structural diversification and neofunctionalization during floral MADS-box gene evolution by C-terminal frameshift mutations. Nucleic Acids Res 2003; 31(15): 4401-9.

[67] Kyozuka J, Harcourt R, Peacock WJ, Dennis ES. *Eucalyptus* has functional equivalents of the *Arabidopsis AP1* gene. Plant Mol Biol 1997; 35(5): 573-84.

[68] Skipper M, Pedersen KB, Johansen LB, *et al.* Identification and quantification of expression levels of three *FRUITFULL*-like MADS-box genes from the orchid *Dendrobium thyrsiflorum* (Reichb. f.). Plant Sci 2005; 169(3): 579-86.

[69] Theissen G, Becker A, Di Rosa A, *et al.* A short history of MADS-box genes in plants. Plant Mol Biol 2000; 42(1): 115-49.

[70] Ma H, Yanofsky MF, Meyerowitz EM. *AGL1-AGL6*, an *Arabidopsis* gene family with similarity to floral homeotic and transcription factor genes. Genes Dev 1991; 5(3): 484-95.

[71] Mena M, Mandel MA, Lerner DR, Yanofsky MF, Schmidt RJ. A characterization of the MADS-box gene family in maize. Plant J 1995; 8(6): 845-54.

[72] Tandre K, Albert VA, Sundas A, Engstrom P. Conifer homologues to genes that control floral development in angiosperms. Plant Mol Biol 1995; 27(1): 69-78.

[73] Shindo S, Ito M, Ueda K, Kato M, Hasebe M. Characterization of MADS genes in the gymnosperm *Gnetum parvifolium* and its implication on the evolution of reproductive organs in seed plants. Evol Dev 1999; 1(3): 180-90.

[74] Purugganan MD, Rounsley SD, Schmidt RJ, Yanofsky MF. Molecular evolution of flower development: diversification of the plant MADS-box regulatory gene family. Genetics 1995; 140(1): 345-56.

[75] Yu H, Goh CJ. Identification and characterization of three orchid MADS-box genes of the *AP1/AGL9* subfamily during floral transition. Plant Physiol 2000; 123(4): 1325-36.

[76] Kaufmann K, Muino JM, Jauregui R, *et al.* Target genes of the MADS transcription factor *SEPALLATA3*: integration of developmental and hormonal pathways in the *Arabidopsis* flower. PLoS Biol 2009; 7(4): e1000090.

[77] Deng W, Ying H, Helliwell CA, *et al. FLOWERING LOCUS C* (*FLC*) regulates development pathways throughout the life cycle of *Arabidopsis*. Proc Natl Acad Sci U S A 2011; 108(16): 6680-5.

[78] Mandel MA, Yanofsky MF. The *Arabidopsis AGL8* MADS box gene is expressed in inflorescence meristems and is negatively regulated by *APETALA1*. Plant Cell 1995; 7(11): 1763-71.

[79] Menzel G, Apel K, Melzer S. Identification of two MADS box genes that are expressed in the apical meristem of the long-day plant *Sinapis alba* in transition to flowering. Plant J 1996; 9(3): 399-408.

[80] Bonhomme F, Sommer H, Bernier G, Jacqmard A. Characterization of *SaMADS D* from *Sinapis alba* suggests a dual function of the gene: in inflorescence development and floral organogenesis. Plant Mol Biol 1997; 34(4): 573-82.

[81] Sung SK, Yu GH, An G. Characterization of *MdMADS2*, a member of the *SQUAMOSA* subfamily of genes, in apple. Plant Physiol 1999; 120(4): 969-78.

[82] Xu Y, Teo LL, Zhou J, Kumar PP, Yu H. Floral organ identity genes in the orchid *Dendrobium crumenatum*. Plant J 2006; 46(1): 54-68.

[83] Mandel MA, Gustafson-Brown C, Savidge B, Yanofsky MF. Molecular characterization of the *Arabidopsis* floral homeotic gene *APETALA1*. Nature 1992; 360(6401): 273-7.

[84] Kempin SA, Savidge B, Yanofsky MF. Molecular basis of the cauliflower phenotype in *Arabidopsis*. Science 1995; 267(5197): 522-5.

[85] Flanagan CA, Ma H. Spatially and temporally regulated expression of the MADS-box gene *AGL2* in wild-type and mutant *Arabidopsis* flowers. Plant Mol Biol 1994; 26(2): 581-95.

[86] Savidge B, Rounsley SD, Yanofsky MF. Temporal relationship between the transcription of two *Arabidopsis* MADS box genes and the floral organ identity genes. Plant Cell 1995; 7(6): 721-33.

[87] Mandel MA, Yanofsky MF. The *Arabidopsis AGL9* MADS box gene is expressed in young flower primordia. Sex Plant Reprod 1998; 11(1): 22-28.

[88] Yu H, Yang SH, Goh CJ. Spatial and temporal expression of the orchid floral homeotic gene *DOMADS1* is mediated by its upstream regulatory regions. Plant Mol Biol 2002; 49(2): 225-37.

[89] Schwarz-Sommer Z, Hue I, Huijser P, *et al.* Characterization of the *Antirrhinum* floral homeotic MADS-box gene *deficiens*: evidence for DNA binding and autoregulation of its persistent expression throughout flower development. Embo J 1992; 11(1): 251-63.

[90] Riechmann JL, Krizek BA, Meyerowitz EM. Dimerization specificity of *Arabidopsis* MADS domain homeotic proteins APETALA1, APETALA3, PISTILLATA, and AGAMOUS. Proc Natl Acad Sci U S A 1996; 93(10): 4793-8.

[91] Hill TA, Day CD, Zondlo SC, Thackeray AG, Irish VF. Discrete spatial and temporal *cis*-acting elements regulate transcription of the *Arabidopsis* floral homeotic gene *APETALA3*. Development 1998; 125(9): 1711-21.

[92] Tilly JJ, Allen DW, Jack T. The CArG boxes in the promoter of the *Arabidopsis* floral organ identity gene *APETALA3* mediate diverse regulatory effects. Development 1998; 125(9): 1647-57.

[93] Yu H, Yang SH, Goh CJ. *DOH1*, a class 1 *knox* gene, is required for maintenance of the basic plant architecture and floral transition in orchid. Plant Cell 2000; 12(11): 2143-60.

[94] Kramer EM, Dorit RL, Irish VF. Molecular evolution of genes controlling petal and stamen development: duplication and divergence within the *APETALA3* and *PISTILLATA* MADS-box gene lineages. Genetics 1998; 149(2): 765-83.

[95] Kramer EM, Jaramillo MA, Di Stilio VS. Patterns of gene duplication and functional evolution during the diversification of the *AGAMOUS* subfamily of MADS box genes in angiosperms. Genetics 2004; 166(2): 1011-23.

[96] Kramer EM, Su HJ, Wu CC, Hu JM. A simplified explanation for the frameshift mutation that created a novel C-terminal motif in the *APETALA3* gene lineage. BMC Evol Biol 2006; 6: 30.

[97] Chen D, Guo B, Hexige S, *et al.* *SQUA*-like genes in the orchid *Phalaenopsis* are expressed in both vegetative and reproductive tissues. Planta 2007; 226(2): 369-80.

[98] Ferrandiz C, Gu Q, Martienssen R, Yanofsky MF. Redundant regulation of meristem identity and plant architecture by *FRUITFULL*, *APETALA1* and *CAULIFLOWER*. Development 2000; 127(4): 725-34.

[99] Fornara F, Parenicova L, Falasca G, *et al.* Functional characterization of *OsMADS18*, a member of the *AP1/SQUA* subfamily of MADS box genes. Plant Physiol 2004; 135(4): 2207-19.

[100] Gocal GF, King RW, Blundell CA, *et al.* Evolution of floral meristem identity genes. Analysis of *Lolium temulentum* genes related to *APETALA1* and *LEAFY* of *Arabidopsis*. Plant Physiol 2001; 125(4): 1788-801.

[101] Gu Q, Ferrandiz C, Yanofsky MF, Martienssen R. The *FRUITFULL* MADS-box gene mediates cell differentiation during *Arabidopsis* fruit development. Development 1998; 125(8): 1509-17.

[102] Immink RG, Hannapel DJ, Ferrario S, *et al.* A petunia MADS box gene involved in the transition from vegetative to reproductive development. Development 1999; 126(22): 5117-26.

[103] Huijser P, Klein J, Lonnig WE, *et al.* Bracteomania, an inflorescence anomaly, is caused by the loss of function of the MADS-box gene *squamosa* in *Antirrhinum majus*. Embo J 1992; 11(4): 1239-49.

[104] Irish VF, Sussex IM. Function of the *apetala-1* gene during *Arabidopsis* floral development. Plant Cell 1990; 2(8): 741-53.

[105] Gustafson-Brown C, Savidge B, Yanofsky MF. Regulation of the *Arabidopsis* floral homeotic gene *APETALA1*. Cell 1994; 76(1): 131-43.

[106] Berbel A, Navarro C, Ferrandiz C, *et al.* Analysis of *PEAM4*, the pea *AP1* functional homologue, supports a model for *AP1*-like genes controlling both floral meristem and floral organ identity in different plant species. Plant J 2001; 25(4): 441-51.

[107] Pan ZJ, Chen YY, Du JS, *et al.* Flower development of *Phalaenopsis* orchid involves functionally divergent *SEPALLATA*-like genes. New Phytol 2014; 202(3): 1024-42.

[108] Hsu HF, Huang CH, Chou LT, Yang CH. Ectopic expression of an orchid (*Oncidium* Gower Ramsey) *AGL6*-like gene promotes flowering by activating flowering time genes in *Arabidopsis thaliana*. Plant Cell Physiol 2003; 44(8): 783-94.

[109] Mandel MA, Yanofsky MF. A gene triggering flower formation in *Arabidopsis*. Nature 1995; 377(6549): 522-4.

[110] Pelaz S, Gustafson-Brown C, Kohalmi SE, Crosby WL, Yanofsky MF. APETALA1 and SEPALLATA3 interact to promote flower development. Plant J 2001; 26(4): 385-94.

[111] Honma T, Goto K. Complexes of MADS-box proteins are sufficient to convert leaves into floral organs. Nature 2001; 409(6819): 525-9.

[112] Hsu HF, Yang CH. An orchid (*Oncidium* Gower Ramsey) *AP3*-like MADS gene regulates floral formation and initiation. Plant Cell Physiol 2002; 43(10): 1198-209.

[113] Hsu HF, Hsieh WP, Chen MK, Chang YY, Yang CH. C/D class MADS box genes from two monocots, orchid (*Oncidium* Gower Ramsey) and lily (*Lilium longiflorum*), exhibit different effects on floral transition and formation in *Arabidopsis thaliana*. Plant Cell Physiol 2010; 51(6): 1029-45.

[114] Chang YY, Chiu YF, Wu JW, Yang CH. Four orchid (*Oncidium* Gower Ramsey) *AP1/AGL9*-like MADS box genes show novel expression patterns and cause different effects on floral transition and formation in *Arabidopsis thaliana*. Plant Cell Physiol 2009; 50(8): 1425-38.

[115] Angenent GC, Busscher M, Franken J, Mol JN, van Tunen AJ. Differential expression of two MADS box genes in wild-type and mutant petunia flowers. Plant Cell 1992; 4(8): 983-93.

[116] Angenent GC, Franken J, Busscher M, Weiss D, van Tunen AJ. Co-suppression of the petunia homeotic gene *fbp2* affects the identity of the generative meristem. Plant J 1994; 5(1): 33-44.

[117] Pnueli L, Hareven D, Broday L, Hurwitz C, Lifschitz E. The *TM5* MADS box gene mediates organ differentiation in the three inner whorls of tomato flowers. Plant Cell 1994; 6(2): 175-86.

[118] Pelaz S, Ditta GS, Baumann E, Wisman E, Yanofsky MF. B and C floral organ identity functions require *SEPALLATA* MADS-box genes. Nature 2000; 405(6783): 200-3.

[119] Tzeng TY, Hsiao CC, Chi PJ, Yang CH. Two lily *SEPALLATA*-like genes cause different effects on floral formation and floral transition in *Arabidopsis*. Plant Physiol 2003; 133(3): 1091-101.

[120] Moon YH, Kang HG, Jung JY, *et al.* Determination of the motif responsible for interaction between the rice APETALA1/AGAMOUS-LIKE9 family proteins using a yeast two-hybrid system. Plant Physiol 1999; 120(4): 1193-204.

[121] Kyozuka J, Kobayashi T, Morita M, Shimamoto K. Spatially and temporally regulated expression of rice MADS box genes with similarity to *Arabidopsis* class A, B and C genes. Plant Cell Physiol 2000; 41(6): 710-8.

[122] Masiero S, Imbriano C, Ravasio F, *et al.* Ternary complex formation between MADS-box transcription factors and the histone fold protein NF-YB. J Biol Chem 2002; 277(29): 26429-35.

[123] Chen MK, Lin IC, Yang CH. Functional analysis of three lily (*Lilium longiflorum*) *APETALA1*-like MADS box genes in regulating floral transition and formation. Plant Cell Physiol 2008; 49(5): 704-17.

[124] Zhang P, Tan HT, Pwee KH, Kumar PP. Conservation of class C function of floral organ development during 300 million years of evolution from gymnosperms to angiosperms. Plant J 2004; 37(4): 566-77.

[125] Skipper M, Johansen LB, Pedersen KB, Frederiksen S, Johansen BB. Cloning and transcription analysis of an *AGAMOUS*- and *SEEDSTICK* ortholog in the orchid *Dendrobium thyrsiflorum* (Reichb. f.). Gene 2006; 366(2): 266-74.

[126] Lopez-Dee ZP, Wittich P, Enrico Pe M, *et al.* *OsMADS13*, a novel rice MADS-box gene expressed during ovule development. Dev Genet 1999; 25(3): 237-44.

[127] Xu HY, Li XG, Li QZ, *et al.* Characterization of *HoMADS 1* and its induction by plant hormones during *in vitro* ovule development in *Hyacinthus orientalis* L. Plant Mol Biol 2004; 55(2): 209-20.

[128] Tzeng TY, Chen HY, Yang CH. Ectopic expression of carpel-specific MADS box genes from lily and lisianthus causes similar homeotic conversion of sepal and petal in *Arabidopsis*. Plant Physiol 2002; 130(4): 1827-36.

[129] Schmidt RJ, Veit B, Mandel MA, *et al.* Identification and molecular characterization of *ZAG1*, the maize homolog of the *Arabidopsis* floral homeotic gene *AGAMOUS*. Plant Cell 1993; 5(7): 729-37.

[130] Pinyopich A, Ditta GS, Savidge B, *et al.* Assessing the redundancy of MADS-box genes during carpel and ovule development. Nature 2003; 424(6944): 85-8.

[131] Rigola D, Pe ME, Fabrizio C, Me G, Sari-Gorla M. *CaMADS1*, a MADS box gene expressed in the carpel of hazelnut. Plant Mol Biol 1998; 38(6): 1147-60.

[132] Kang HG, Noh YS, Chung YY, *et al.* Phenotypic alterations of petal and sepal by ectopic expression of a rice MADS box gene in tobacco. Plant Mol Biol 1995; 29(1): 1-10.

[133] Rutledge R, Regan S, Nicolas O, *et al.* Characterization of an *AGAMOUS* homologue from the conifer black spruce (*Picea mariana*) that produces floral homeotic conversions when expressed in *Arabidopsis*. Plant J 1998; 15(5): 625-34.

[134] Angenent GC, Franken J, Busscher M, *et al.* A novel class of MADS box genes is involved in ovule development in petunia. Plant Cell 1995; 7(10): 1569-82.

[135] Colombo L, Franken J, Koetje E, *et al.* The petunia MADS box gene *FBP11* determines ovule identity. Plant Cell 1995; 7(11): 1859-68.

[136] Chen YY, Lee PF, Hsiao YY, *et al.* C- and D-class MADS-box genes from *Phalaenopsis equestris* (Orchidaceae) display functions in gynostemium and ovule development. Plant Cell Physiol 2012; 53(6): 1053-67.

[137] Salemme M, Sica M, Gaudio L, Aceto S. The *OitaAG* and *OitaSTK* genes of the orchid *Orchis italica*: a comparative analysis with other C- and D-class MADS-box genes. Mol Biol Rep 2013; 40(5): 3523-35.

[138] Wang SY, Lee PF, Lee YI, *et al.* Duplicated C-class MADS-box genes reveal distinct roles in gynostemium development in *Cymbidium ensifolium* (Orchidaceae). Plant Cell Physiol 2011; 52(3): 563-77.

[139] Zahn LM, Leebens-Mack J, DePamphilis CW, Ma H, Theissen G. To B or Not to B a flower: the role of *DEFICIENS* and *GLOBOSA* orthologs in the evolution of the angiosperms. J Hered 2005; 96(3): 225-40.

[140] Tsai WC, Kuoh CS, Chuang MH, Chen WH, Chen HH. Four *DEF*-like MADS box genes displayed distinct floral morphogenetic roles in *Phalaenopsis* orchid. Plant Cell Physiol 2004; 45(7): 831-44.

[141] Hsu CC, Wu PS, Chen TC, *et al.* Histone acetylation accompanied with promoter sequences displaying differential expression profiles of B-class MADS-box genes for *Phalaenopsis* floral morphogenesis. PLoS One 2014; 9(12): e106033.

[142] Tsai WC, Lee PF, Chen HI, *et al. PeMADS6*, a *GLOBOSA/PISTILLATA*-like gene in *Phalaenopsis equestris* involved in petaloid formation, and correlated with flower longevity and ovary development. Plant Cell Physiol 2005; 46(7): 1125-39.

[143] Tsai WC, Pan ZJ, Hsiao YY, *et al.* Interactions of B-class complex proteins involved in tepal development in *Phalaenopsis* orchid. Plant Cell Physiol 2008; 49(5): 814-24.

[144] Kim S-Y, Yun P-Y, Fukuda T, *et al.* Expression of a *DEFICIENS*-like gene correlates with the differentiation between sepal and petal in the orchid, *Habenaria radiata* (Orchidaceae). Plant Sci 2007; 172(2): 319-26.

[145] Chang YY, Kao NH, Li JY, *et al.* Characterization of the possible roles for B class MADS box genes in regulation of perianth formation in orchid. Plant Physiol 2010; 152(2): 837-53.

[146] Yu D, Kotilainen M, Pollanen E, *et al.* Organ identity genes and modified patterns of flower development in *Gerbera hybrida* (Asteraceae). Plant J 1999; 17(1): 51-62.

[147] Aceto S, Montieri S, Sica M, Gaudio L. Molecular evolution of the *OrcPI* locus in natural populations of Mediterranean orchids. Gene 2007; 392(1-2): 299-305.

[148] Cantone C, Sica M, Gaudio L, Aceto S. The *OrcPI* locus: Genomic organization, expression pattern, and noncoding regions variability in *Orchis italica* (Orchidaceae) and related species. Gene 2009; 434(1-2): 9-15.

[149] Aceto S, Cantone C, Chiaiese P, *et al.* Isolation and phylogenetic footprinting analysis of the 5'-regulatory region of the floral homeotic gene *OrcPI* from *Orchis italica* (Orchidaceae). J Hered 2010; 101(1): 124-31.

[150] Pan ZJ, Cheng CC, Tsai WC, *et al.* The duplicated B-class MADS-box genes display dualistic characters in orchid floral organ identity and growth. Plant Cell Physiol 2011; 52(9): 1515-31.

[151] Cantone C, Gaudio L, Aceto S. The *PI/GLO*-like locus in orchids: duplication and purifying selection at synonymous sites within Orchidinae (Orchidaceae). Gene 2011; 481(1): 48-55.

[152] Salemme M, Sica M, Gaudio L, Aceto S. Expression pattern of two paralogs of the *PI/GLO*-like locus during *Orchis italica* (Orchidaceae, Orchidinae) flower development. Dev Genes Evol 2011; 221(4): 241-6.

[153] Mondragon-Palomino M, Hiese L, Harter A, Koch MA, Theissen G. Positive selection and ancient duplications in the evolution of class B floral homeotic genes of orchids and grasses. BMC Evol Biol 2009; 9: 81.

[154] Kanno A, Saeki H, Kameya T, Saedler H, Theissen G. Heterotopic expression of class B floral homeotic genes supports a modified ABC model for tulip (*Tulipa gesneriana*). Plant Mol Biol 2003; 52(4): 831-41.

[155] Nakamura T, Fukuda T, Nakano M, *et al.* The modified ABC model explains the development of the petaloid perianth of *Agapanthus praecox* ssp. *orientalis* (Agapanthaceae) flowers. Plant Mol Biol 2005; 58(3): 435-45.

[156] Mondragon-Palomino M, Theissen G. MADS about the evolution of orchid flowers. Trends Plant Sci 2008; 13(2): 51-9.

[157] Mondragon-Palomino M, Theissen G. Conserved differential expression of paralogous *DEFICIENS* and *GLOBOSA*-like MADS-box genes in the flowers of Orchidaceae: refining the 'orchid code'. Plant J 2011; 66(6): 1008-19.

CHAPTER 7

Perspectives on Systematic Analyses of Gene Function in *Arabidopsis thaliana*: New Tools, Topics and Trends

Cordelia Bolle, Anja Schneider and Dario Leister*

Lehrstuhl für Molekularbiologie der Pflanzen (Botanik), Department Biologie I, Ludwig-Maximilians-Universität München, Germany

Abstract: Since the sequencing of the nuclear genome of Arabidopsis thaliana more than ten years ago, various large-scale analyses of gene function have been performed in this model species. In particular, the availability of collections of lines harbouring random T-DNA or transposon insertions, which include mutants for almost all of the ~28,000 A. thaliana genes, has been crucial for the success of forward and reverse genetic approaches. In the foreseeable future, genome-wide phenotypic data from mutant analyses will become available for Arabidopsis, and will stimulate a flood of novel in-depth gene-function analyses. In this review, we consider the present status of resources and concepts for systematic studies of gene function in A. thaliana. Current perspectives on the utility of loss-of-function and gain-of-function mutants will be discussed in light of the genetic and functional redundancy of many A. thaliana genes.

Keywords: *Arabidopsis thaliana*, EMS, forward genetics, gain of function, loss of function, mutant, mutation, phenotype, reverse genetics, transposon, T-DNA.

INTRODUCTION

Because of its small size, short life cycle and compact genome, *Arabidopsis thaliana* (hereafter *Arabidopsis*) has been developed into a model system for plants. Since its nuclear genome was completely sequenced more than 10 years ago [1], the quest to decipher the functions of all its genes has been a major strand in *Arabidopsis* research. Forward (from-phenotype-to-gene) and reverse (from-gene-to-phenotype) genetics have been the main workhorses *en route* to this goal. Even before the *Arabidopsis* Genome Initiative got underway, numerous forward genetic screens had identified mutants with abnormal phenotypes that resulted in the identification and functional characterization of many genes for specific plant

***Corresponding author Dario Leister:** Lehrstuhl für Molekularbiologie der Pflanzen (Botanik), Department Biologie I, Ludwig-Maximilians-Universität München, Großhaderner Str. 2, D-82152 Planegg-Martinsried, Germany; Tel: +49-89/2180 74550; Fax: +49-89/2180 74599; E-mail: leister@lmu.de

Atta-ur-Rahman (Ed.)

functions. Identification of the mutated genes is usually the most labour-intensive step in classical forward genetics. With increasing knowledge of the sequences of *Arabidopsis* genes and their transcripts, reverse genetic screens have become possible, which complement the classical approach by permitting direct analysis of mutants for specific genes, selected for their potential impact on a biological function of interest. According to the Annual Report 2011 of The Multinational Coordinated *Arabidopsis thaliana* Functional Genomics Project (http://www.arabidopsis.org/portals/masc/masc_docs/masc_reports.jsp), the experimentally verified functions of just over 9600 of the ~28,000 *Arabidopsis* genes had been annotated as of March 2011.

In silico approaches can assess the functions of *Arabidopsis* genes whose homologs have been characterized in other species, or which code for proteins with conserved domains, and gene functions can be predicted by correlating data derived from functional genomics analyses, such as large-scale mRNA, protein or metabolite profiling [2-4]. However, to decipher the biological function(s) of every gene in *Arabidopsis* will require much further effort, and many projects have been initiated to create the resources required for this formidable task [5]. Technological advances have led to the assembly of large collections of *Arabidopsis* mutants bearing molecularly mapped insertions. Besides total inactivation of the gene function, some of the newly developed mutagenesis techniques allow one to alter the dosage or function of gene products, making refined analyses of gene-function relationships possible. Several databases are now available that permit researchers to scan the *Arabidopsis* genome for mutations in genes of interest and obtain the corresponding mutant lines from public stock centres.

To elucidate the function of the entire complement of *Arabidopsis* genes requires ways to systematically assess a wide range of phenotypes, which can be applied in a high-throughput manner to large numbers of mutants. Saturated collections of loss-of-function and gain-of-function lines will also be very valuable. A further challenge arises from the recognition that the vast majority of gene knock-outs in *Arabidopsis* do not give rise to obvious phenotypes, and might require functional characterisation by –omics-type analyses or the simultaneous inactivation/down-regulation of more than one gene. In this review, we summarize the state of the art in systematic gene-function analyses and highlight tools and trends in the application of systematic forward and reverse genetics to *Arabidopsis*.

TOOLS FOR THE ANALYSIS OF GENE FUNCTIONS

Several methods have been developed that enable one to change the amount or the nature of gene products either by altering the original gene through the introduction of mutations or by introducing transgenes. Since the *Arabidopsis* genome sequence was published most efforts have focused on generating loss-of-function lines for reverse genetics. In addition, methods for overexpressing genes have gained in importance more recently [6, 7].

Loss-of-Function Mutants (Knock-Out Approach)

The most straightforward approach to investigating the function of a gene is to characterize the phenotypic changes associated with its total inactivation [8]. Loss of gene function can be achieved by introducing point mutations or short insertions/deletions, typically by means of chemical or physical mutagenesis, or by disrupting genes by inserting larger DNA sequences, such as T-DNAs or transposable elements.

Point Mutation and Short Insertions/Deletions

Point mutations can be introduced with high frequency by mutagenization with chemicals such as ethyl methanesulfonate (EMS), but only a small fraction of such mutations actually lead to total loss of gene function, *e.g.* when they generate a premature stop codon or alter functionally critical amino acids [9]. Similarly, short insertions/deletions usually eliminate gene function only if the coding sequence is affected and a frameshift results. Point mutations and short insertions/deletions have the disadvantage that the location of the mutation is random and cannot be predetermined. Therefore, in order to have a reasonable probability of finding loss-of-function mutations for a given gene, large numbers of mutants must be generated. To remove the unwanted background mutations, several rounds of backcrossing to the wild type are usually required.

The identification of individual, randomly induced point mutations or short insertions/deletions within the genome of a mutant is a laborious process involving map-based cloning [10], although recent advances in high-throughput sequencing technology [10, 11] might make resequencing of the entire mutant genome the method of choice in future. Because *Arabidopsis* reference genome sequences (such as the Col-0 ecotype) are already available, new generation sequencing of mutant genomes will become more and more time and cost effective. Another approach to detect small changes within the genome is an adaptation of the microarray technology, which detects single nucleotide polymorphisms (SNPs) through hybridizing genomic DNA to oligonucleotides representing the entire genome spotted on chips [12]. With continuous advances in sequencing technologies, genome-based methods can replace the conventional marker-based genotyping approach. Nevertheless, because EMS-mutated plants contain many additional point mutations, it might be recommendable to narrow down first the mutation to a chromosome arm with map-based approaches and conventional techniques.

For reverse genetic approaches employing EMS-induced mutations, the "Targeting Induced Local Lesions IN Genomes" (TILLING) method has been developed [13, 14]. Here, the endonuclease CEL I identifies mismatches in heteroduplexes between wild-type and mutant DNA sequences after PCR-based amplification of genomc sequences of interest [15]. For systematic reverse genetics screens, each target coding sequences must be covered by PCR amplicons [16]. 9550 TILLING lines in Col-0, in which the mutation has been assigned to a specific gene have been donated to public stock centres (Table **1**). Additionally, to optimize for mutation selection, tetraploid Col-0 was used to generate a collection at UC Davis TILLING Core [17]. Furthermore, two other accessions, C24 and L*er*, were used to generate EMS mutagenized collections [18, 19]. Most point mutations do not cause complete gene knock-out, but rather a decrease in gene expression (see also Section 2.2) or a change in the function of the gene product due to alterations in its sequence. Amino acid exchanges can lead to hypomorphic alleles displaying a range of phenotypes, allowing the analysis of important domains of the protein. Such variants can be pinpointed by searching the TILLING database (Table **2**). Variation in naturally occurring *Arabidopsis* accessions is a helpful resource with which relevant point mutations or other genome modifications can be detected. The release of the genome sequences of over 1100 different accessions (as of September 2014) within the "1001 genomes project" makes this an important tool for understanding natural variation of sequences and correlating them with evolutionary and adaptive processes [20-22] (Table **2**).

Insertional Mutagenesis

Insertional mutagenesis has two major advantages: the mutations are labelled by the inserted fragment of known sequence ('tag') and insertions within the coding region have a high probability of eliminating the gene function. In *Arabidopsis*, T-DNA or transposons -- mainly the *Activator* (*Ac*)/*Dissociation* (*Ds*) system -- are usually employed as insertional mutagens [48]. In an international effort involving many laboratories using both systems, a large number of insertion mutants has been generated (Table **1**) which now provide nearly genome-wide coverage -- about 96% of all *Arabidopsis* genes according to the Annual Report 2011 of the Multinational Coordinated *Arabidopsis thaliana* Functional Genomics Project (http://www.arabidopsis.org/portals/masc/masc_docs/masc_reports.jsp).

Table 1. Resources of lines for gene function analyses in *Arabidopsis*.

Collection	Type	Genetic Background	Number of Lines	Number of Insertions Mapped to Genome	Refs.
Loss-of-function or knock-down					
SALK lines (SALK Institute, USA)	T-DNA	Col-0	137,259, of which 38,333 are homozygous (in progress)	166,127; in homozygous lines 50,090	[23]
SAIL lines (Syngenta, USA)	T-DNA	Col-0, Col-3	54,612, of which 5,124 are homozygous	62,041	[24]
Wisconsin lines	Ds-Lox	Col-0	32,614, 2,055 homozygous	21,406	[25]
GABI-KAT lines (GABI, Germany), closed 2014	T-DNA	Col-0	130,654	98,559	[26]
JIC SM (SLAT collection)	En/Spm	Col	~48,000; 24,000 single seed lines	26,007	[27]
CSHL-lines (Martienssen Transposon Insertion lines) (either gene or enhancer trap)	Ds-GUS	*Ler*	24,444	27,398	[28-30]
RIKEN transposon-tagged lines	Ds	No-0	18,622 lines (in progress)	18,622 lines (in progress)	[31-35]
INRA-Versailles lines (FLAGdb)	T-DNA with GUS	Ws	~55,000	37,142	[36, 37]
JIC SM FST	dSpm	*Ler*	22,664	24,206	[27]
AGRIKOLA	RNAi	Col	22,478 constructs, of which 2,519 are transformed into plants	-	[38]
amiRNA (CSHL-based 2010 project)	amiRNA		17,699 clones		see Table **2**
Seattle *Arabidopsis* TILLING Project (closed)	EMS	Col-0 *erecta*	9,550		see Table **2**
Gain-of-function					
FOX hunting lines	fl cDNAover-expression	Col	22,000	-	[39]
Weigel lines	T-DNA$_{AT}$	Col-7	16,398 in pools		[40]
RIKEN activation-tagged lines	T-DNA$_{AT}$	Col	55,000	1,172 (in progress)	[41, 42]
SK (Saskatoon) lines	T-DNA$_{AT}$	Col-4	~70,000	15,507	[43]
TAMARA	En/Spm$_{AT}$	Col-0	9,471 lines	-	[44]
CRES-T	chimeric repressors		395 (in progress)	-	[45]
Chromatin charting	luciferase	Col-0	277 (in progress)	in progress	[46], [47]

Collections with less then 1000 lines are not listed unless specifically referenced in the main text. Numbers of publicly available lines have been obtained from the corresponding websites/references, the stock centers NASC, ABRC, RIKEN Bioresource center and http://signal.salk.edu/Source/AtTOME_Data_Source.html (as of April 2014). T-DNA$_{AT}$, T-DNA activiation tagged; En/Spm$_{AT}$, En/Spm activation-tagged.

The regions flanking the insertions (flanking sequence tags, FST) in over 452,000 lines have been sequenced by several laboratories and mapped to the *Arabidopsis* reference genome to identify the precise insertion sites. Recently, next generation sequencing was established [49] to identify T-DNA insertions by the "Salk Group" in existing SALK, GABI, SAIL and Wisconsin lines. By March 2015 about 313,200 insertions had been identified and linked to about 186,300 T-DNA lines, confirming that many T-DNA lines contain more than one insertion. This additional information will enhance the precision with which the insertion site can be mapped. All these insertion lines have been indexed and deposited in public stock centres. The FSTs and T-DNASeq data are accessible for Internet-based searches on (Table **2,** [50]).

Because stock centres distribute T-DNA or transposon lines usually as populations that segregate for the mutation, a genotyping step is necessary to identify homozygous plants for the analysis of phenotypes [51]. Recently, populations of homozygous SALK lines were generated, which can be used directly for phenotypic screens [52]. To date (April 2015) these homozygous lines represent 50,090 mapped insertions. However, a certain proportion of hemizygous lines escaped systematic genotyping, so that a certain fraction -- perhaps as large as 20% -- might still not be homozygous (our unpublished results). Moreover, because T-DNA lines often contain more than one insertion, for the unambiguous assignment of phenotypes to mutations either at least two insertion alleles for each gene should be analysed and/or the mutation be complemented with the wild-type gene.

Mutants with Reduced Gene Expression (Knock-Down)

The utility of loss-of-function lines is limited when mutations result in lethality, or in cases of genetic redundancy, when more than one copy of the gene exists as a result of tandem or segmental duplications or the evolution of multigene families. To overcome these drawbacks, transgene-mediated gene silencing can be used to reduce but not completely abolish ("knock-down") the expression of gene(s) of interest. Such "knock-down" lines allow one simultaneously to down-regulate several, sequence-related genes and thus avoid the complications associated with genetic redundancy. Because knock-down lines are not null mutants and show a diverse pattern of phenotypes due to differences in the levels of activity remaining, the more subtle phenotypes induced by the partial or conditional inactivation of essential genes can also be studied.

Table 2. Databases/websites for gene function analyses.

Resource/database	Content	Location
Stock Centers and general information		
The *Arabidopsis* Information Resource (TAIR)	Information, tools, databases	http://www.arabidopsis.org/
Araport	Information, tools, databases	https://www.araport.org/
Arabidopsis Biological Resource Center (ABRC)	Information/stock centre	http://abrc.osu.edu/
European *Arabidopsis* Stock Centre (NASC)	Stock centre	http://arabidopsis.info/
Riken BioResource Center	Stock Centre (Japan)	http://epd.brc.riken.jp/en/arabidopsis
Arabidopsis thaliana Resource Centre for Genomics	Databases, stock centre	http://publiclines.versailles.inra.fr/
Databases with annotated FSTs		
T-DNA Express		http://signal.salk.edu/cgi-bin/tdnaexpress
SeqViewer		http://tairvm09.tacc.utexas.edu/servlets/sv
SimpleSearch (GABI-KAT)		http://www.gabi-kat.de/simplesearch.html
Collection of lines and/or phenotypes		
SALK T-DNA collection	Information/databases on SALK T-DNA lines	http://signal.salk.edu/tabout.html
GABI-KAT lines	Information/Database for GABI-KAT lines	http://www.gabi-kat.de/
GABI-DUPLO lines	Information on genetically unlinked gene pairs in *Arabidopsis* and double mutant lines	http://www.gabi-kat.de/duplo.html
Arabidopsis Genetrap Website at Cold Spring Harbor Lab	GUS patterns, phenotypes and mapping data of transposon insertions	http://genetrap.cshl.edu/
RIKEN transposon- and activation-tagged lines	Information on lines and RIKEN *Arabidopsis* Phenome Information Database (RAPID) with phenotypes of 4000 lines	http://pfgweb.gsc.riken.go.jp/index.html http://epd.brc.riken.jp/en/seed/transp http://rarge.psc.riken.jp/dsmutant/:index.pl http://rarge.psc.riken.jp/phenome/

Table 2: contd...

FOX hunting lines	Project description	http://pfgweb.gsc.riken.go.jp/fh-line.html
INRA Versailles lines	Information on INRA lines, phenotypes in Agrobact+ database	http://publiclines.versailles.inra.fr/
SK (Saskatoon) lines	Project description and sequence information	http://aafc-aac.usask.ca/FST/
Seattle *Arabidopsis* TILLING Project	Information on how to find point mutations in gene-of-interest	http://tilling.fhcrc.org/
UC Davis TILLING Core	Information on diploid and tetraploid tilling	http://tilling.ucdavis.edu/index.php/Arabidopsis_TILLING
Arabidopsis Genomic RNAi Knock-out Line Analysis (AGRIKOLA)	RNAi lines	http://www.agrikola.org
amiRNA (CSHL-based 2010 project)	Project description and vector information, 17,699 clones available (April 2015)	http://2010.cshl.edu/arabidopsis/2010/scripts/main2.pl
fioreDB	Project description and database	http://www.cres-t.org/fiore/public_db/f_help.shtml
Phenopsis DB	Phenotyping Database, combining phenotypic data with the associated climatic conditions of growth	http://bioweb.supagro.inra.fr/phenopsis/
Specialized projects		
Seed Genes Project	List of essential genes	http://www.seedgenes.org/
Chloroplast 2010 project	Project description and database	http://www.plastid.msu.edu/
Chloroplast Function Database	Project description and database	http://rarge.psc.riken.jp/chloroplast/
PhenoLeaf	Project description and database	http://genetics.umh.es/phenoleaf
PlantMetabolomics	Project description and database	http://plantmetabolomics.vrac.iastate.edu/ver2/index.php
1001 Genomes Project	Whole-genome sequence variation in 1001 accessions of *A. thaliana*	http://1001genomes.org/ http://signal.salk.edu/atg1001/index.php

Silencing is normally achieved by post-transcriptional down-regulation of transcript accumulation *via* small RNAs that act in a sequence-specific manner by base pairing to complementary mRNA molecules. Various strategies for small RNA-based gene silencing have been developed [53]. The first to be developed was the antisense approach, in which part of a gene was expressed in reverse orientation, usually under the control of the Cauliflower Mosaic Virus (CaMV) 35S promoter [54]. Later the RNA interference (RNAi) approach was introduced. This uses a binary hairpin RNA vector into which gene-specific tags (GSTs) are cloned [55]. Transformation of *Arabidopsis* with RNAi constructs is being systematically conducted by the AGRIKOLA (*Arabidopsis* Genomic RNAi Knock-Out Line Analysis) Consortium to create a collection of silenced *Arabidopsis* lines which are available through the Nottingham Arabidopsis Stock Centre (NASC) [56] (Tables **1** and **2**). A newer approach involves the use of artificial microRNAs (amiRNAs) [53, 57, 58]. The systematic design of amiRNA has led to the generation of 17,699 clones for use in *Arabidopsis* (as of April 2015), providing an important tool for future functional genomics studies (Tables **1** and **2**). The Chimeric Repressor Silencing Technology (CRES-T) represents a novel method for creating loss-of-function mutations for the analysis of redundant plant transcription factors [59]. With CRES-T, a transcription activator of interest can be fused to a peptide or protein that converts it into a transcriptional repressor, which dominantly suppresses the expression of its target genes even in the presence of the redundant transcription factor. The phenotypic information obtained from the application of CRES-T in *Arabidopsis* is stored in the Web-based interface FioreDB [45] (Tables **1** and **2**).

Gain-of-Function Lines

Gain-of-function lines can be used to dissect the functions of genes that are genetically or functionally redundant, *i.e.* which are members of gene families or are tandemly or segmentally duplicated, or whose function can be compensated for by alternative regulatory pathways [41]. The phenotypes caused by gain-of-function mutations segregate as dominant traits [8, 40] and individual members of a gene family can produce a gain-of-function mutant phenotype that cannot be modified by other members of the family [60, 61]. Often the mutant phenotypes induced by loss-of-function and gain-of-function approaches are complementary to each other. Because of the dominant nature of the gain-of-function phenotypes, there is no need for an additional genotyping step to identify lines homozygous for the transgene.

Overexpression of a gene driven by a strong promoter can result in gain-of-function phenotypes, especially when expression is also ectopic. This strategy requires the use of full-length cDNA clones. The RIKEN *Arabidopsis* full-length (RAFL) collection now makes large numbers of such cDNA clones available. The RAFL collection was generated by inserting full-length cDNAs in the correct orientation between the CaMV 35S promoter and the NOS terminator and verifying their sequences [62]. About 22,000 different RAFL cDNA clones were then pooled and transformed into *Arabidopsis* to generate so-called 'FOX' (Full-length cDNA Over-eXpressing gene) lines [39]. The FOX hunting system is useful for systematic phenotypic analysis of the function of each inserted gene. Visible phenotypes were induced in about 9% of transformed lines and details of these have been collected in a database (see Section 3.2 and Table **2**).

In the activation tagging approach, DNA sequences carrying enhancers (*e.g.* four tandemly arranged copies of the CaMV 35S enhancer) are inserted randomly into the genome [63]. Large numbers of activation-tagging lines have been produced, and phenotypes observed could be attributed to specific genetic events by sequencing the regions flanking insertions and mapping these to the reference genome [27, 40, 41, 43, 44, 63-65]. The distances between insertions and the genes affected were found to vary from 0.4 to 8.2 kb. Enhancers can control the expression of genes on both sides of the insertion and, indeed, in some cases more than one gene is affected [40, 42]. In addition to dominant gain-of-function phenotypes, knock-outs are also observed in these lines, when an insertion is located within a coding region. Between 0.1 and 1 % of the activation tagging lines showed visible phenotypes [27, 40, 41, 44, 64, 65]. A modified activation-tagging method has been developed which uses pEnLOX lines containing four tandemly repeated CaMV 35S transcriptional enhancers flanked by two *loxP* sites, and pCre lines that contain the *CRE* gene [66]. By crossing pENLOX lines to pCre lines, the CaMV 35S enhancers can be deleted, causing reversion to the wild-type phenotype, and allowing for speedier confirmation of gene-function relationships.

The Chromatin Charting Project has set itself the task of producing over 15,000 *Ds*-tagged lines of *Arabidopsis*, each of which contains within the *Ds*-tagging cassette a CaMV 35S::luciferase gene (which allows monitoring of position effects) and a *lac* operator repeat (which allows the tagged loci to be visualized by expressing fusions of fluorescent proteins with LacI within the nuclei of living plants). The selected lines will be used to evaluate transgene position effects and how these are affected by developmental or externally applied cues; since position effects on gene expression arise from alterations in chromatin architecture, this approach provides insights into global epigenetic control [47, 67] (Table **1**).

SYSTEMATIC FORWARD GENETIC SCREENS

Forward genetic screens have traditionally been used to identify genes involved in a specific biological process of interest. A range of mutant collections have been analysed, including chemically or physically mutagenized populations – which have the advantage that it is, in principle, possible to attain saturation, *i.e.* to generate multiple mutant alleles for each gene – and tagged mutants, which allow the mutated gene to be identified in a relatively straightforward manner. Screens for a wide variety of phenotypes have been carried out in *Arabidopsis*, and only a selection can be discussed here.

Generic Screens

The RIKEN Activation Tagged Lines, comprising about 55,000 independently transformed *Arabidopsis* lines [41, 42] (see Table **1**), have been screened for visually discernible phenotypes (Table **2**). Of four dominant mutants that showed hyponastic leaves, downward-pointing flowers and decreased apical dominance, two were caused by activation of the gene *ASYMMETRIC LEAVES2* (*AS2*) and two by activation of *ASL1/LBD36* [41].

The FOX lines (see Section 2.3, Table **1**) were also screened for abnormal morphologies, fertility and leaf coloration. A total of 1487 candidate morphological mutants were found among 15,547 transformants, and of 115 pale green T_1 mutants, 59 lines displayed the mutant phenotypes in more than 50% of the T_2 progeny, suggesting that the phenotypes were caused by gain of function mutations (see Section 2.3). Two leaf coloration mutants were further characterized and one mutant was found to result from up-regulation of *AtPDH1* (*Arabidospsis* prokaryotic DEVH box helicase1); knock-out of this same gene resulted in an albino phenotype [39].

Screens for Developmental Phenotypes

The SeedGenes Project [68-71] was designed to identify *EMBRYO DEFECTIVE* (*EMB*) genes required for embryo or seed development, and as of April 2015 the database SeedGenes (Table **2**) includes information on 481 genes required for seed development and 888 mutant alleles with known disruptions in these genes. *EMB* genes encode proteins involved in basic cellular processes, such as DNA replication, RNA processing and protein synthesis, as well as chloroplast proteins, indicating that a functional chloroplast is needed for normal embryo development in *Arabidopsis*.

Furthermore, over 500 seedling-lethal mutants were identified in a set of 38,000 insertion mutants, of which 54 were molecularly characterized and 22 mutants were described in detail [72]. Many of the mutants displayed altered pigmentation and affect genes encoding proteins predicted to reside in the chloroplast.

Developmental pathways can also be addressed by genetic screens, one example being the genes involved in pollen exine production, which led to the identification of 56 mutants [73]. To isolate mutants defective in early steps of meiotic recombination, a specialized two-step genetic screen employing 55,000 T-DNA insertion lines has been carried out [74]. In the first step, all lines were screened for fertility defects on the basis of reduced silique elongation. In the second step, differential interference contrast microscopy was used to analyse male meiotic products in developing pollen mother cells. Four genes involved in the repair of meiotic DNA double-stranded breaks were identified, together with five genes necessary for formation of meiotic DNA double-strand breaks [74].

Screens for Genes Important for Male or Female Gametophyte Development and Reproduction

The life cycle of plants alternates between a haploid gametophyte and a diploid sporophyte phase. Mutations that eliminate or disrupt the function of the haploid pollen (male gametophyte) or of the embryo sac (female gametophyte) can only be maintained in the heterozygous state, because such mutations cannot be transmitted through the defective gamete (reviewed in [75-77]). In addition, mutations in sporophytically expressed genes might result in male or female sterility [78, 79].

Visual inspection of siliques for 50% or more desiccated ovules, as well as screens for aberrant transmission of an antibiotic resistance marker gene, led to the identification of gametophytic mutants [80], [81-86]. In *Arabidopsis*, due to the widely used "floral dipping" transformation protocol the primary target of T-DNA integration is the embryo sac. Therefore, it can be difficult to recover T-DNA insertional knock-out lines, in which female gametophyte function is severely compromised [81]. To overcome this problem, transposon insertion lines have been used. *Ds* transposable elements are mobilized during the sporopyhtic stage of a plant's life cycle and most *Ds* transposon insertion lines possess further advantages like single-copy insertions without generating DNA rearrangements or truncations [28]. 24,000 *Ds* insertion lines containing either a *Ds* gene trap or an enhancer trap element were screened and 130 *Arabidopsis* mutants with defects in female gametophyte development and function were identified [87]. The inheritance of the *Ds* element could be followed by the linked antibiotic resistance marker. Given the dominant mode of inheritance of the resistance phenotype, a segregation of 3:1 is expected in the progeny of a heterozygous *Ds* insertion line. Aberrant segregation ratios lower than 2:1 are often the result of a gametophytic defect. For 1.38% of the screened lines a segregation ratio of 1.5:1 or less was found and a wide variety of mutant phenotypes were observed, such as defects in different stages of embryo sac development and in processes such as pollen tube guidance, fertilization or early embryo development. Besides female gametophytic mutants, 109 lines showing predominant defects in male gametophyte transmission were also identified in this collection [88].

A genetic screen using 1,666 EMS-mutagenised lines to identify mutations that would induce gynogenesis or androgenesis in *Arabidopsis thaliana* failed, implying that EMS-induced mutations in the male genome are not sufficient to induce gynogenesis or androgenesis in *Arabidopsis,* although a spontaneous gynogenesis could be observed [89].

Screens for Altered Metabolism or Stress Tolerance

In a screen of the TAMARA gain-of-function collection [44] for mutants with enhanced levels of phenolic compounds, up-regulation of an R2R3-MYB transcription factor was found to be responsible for the phenotype [41, 44, 90]. Genes controlling proanthocyanidin biosynthesis have also been identified in both the TAMARA and Saskatoon collections (Table **2**) [43].

A small subset of the FOX lines has also been assessed for mutants affecting stress-inducible transcription factors. Among 43 cDNAs tested in a transgenic plant library, the authors identified salt-tolerant lines, all of which overexpressed the same bZIP-type transcription factor, AtbZIP60 [91].

The COS (controlled cDNA overexpression) system is based on a cDNA expression library that is driven by an inducible promoter. Sets of 20,000 to 40,000 transgenic seeds/seedlings were screened in three different ways, namely for ABA insensitivity, salt tolerance and for activation of a stress-responsive alcohol dehydrogenase-luciferase reporter system [92, 93]. Twenty-seven cDNAs conferring dominant, inducible stress-tolerance phenotypes were identified, and recloning of several of these cDNAs confirmed the observed phenotypes [92].

Systemic acquired resistance (SAR) is a defense mechanism against infection. To isolate novel components of this pathway a high-throughput assay has been efficiently developed, demonstrating that also biotic interactions can be genetically accessed [94].

Chemical Genetic Screens

Treatment of mutant collection with a chemical compound to elucidate gene functions and signal transduction pathways are widely applied [95]. Limitations of this approach can be limited uptake of the compound by the plant or transport over membranes, and that possible modifications such as acetylation might inactive the compound. A prerequisite for these so called "chemical genetic screens" is the availability of a library of chemicals. The chemicals are usually used to screen for changes of a specific physiological parameter. For strigolactones, for example, that play a role during parasitic seed germination of plants such as Striga, a collection of related small molecules -- cotylimides -- that act as genetic suppressors of strigolactone levels were identified this way [96]. These cotylimides were used in a suppressor screen with T-DNA mutants, and light-signaling genes were identified as positive regulators of strigolactone levels [96]. Another example is the compound pyrabactin. Pyrabactin has been isolated as a synthetic seed germination inhibitor that mimics abscisic acid (ABA) in a highly selective way, thereby activating the ABA pathway. *Arabidopsis* EMS mutants were screened with pyrabactin and a new component of the ABA pathway, the *PYRABACTIN RESISTANCE 1* (*PYR1*) gene was identified. It codes for a START protein, which was characterized as the long-elusive ABA receptor [97]. Other screens involved herbicide treatment and the isolation of resistant lines. As such, EMS mutants were treated with paraquat which accepts electrons from photosystem I, and PARAQUAT RESISTANT 1 (PAR1), a Golgi-localized transporter protein, isolated [98].

Besides EMS or insertion mutants multiple *Arabidopsis* natural accessions are a rich source for applying chemical genetic screens. When *Arabidopsis* natural accessions were subjected to a library of more than 10,000 compounds, twelve accession-specific molecules were identified [99]. The natural resistance to hypostatin, an inhibitor of cell expansion, was identified in this way and characterized as HYR1, a UDP glycosyltransferase.

Chemical molecules that inhibit a specific protein function result in a similar phenotype like corresponding loss-of-function mutants. *Vice versa*, plants treated with a compound that activates a protein function resemble gain-of-function mutants. Variation in concentration of the chemical compound can act similar to allelic series of mutations. If a chemical is able to interfere with the function of all members of a protein family, it can substitute for mutants of a higher order. The ability to perturb protein functions only in the presence of a chemical and therefore for a limited time allows the study of essential genes that would display a lethal phenotype if permanently mutated. Nevertheless, currently only a limited number of chemical compounds with a defined function are available to be applied for biological questions. Helpful are databases such as ChemMine [100] or Pubchem (http://pubchem.ncbi.nlm.nih.gov/).

Non-Invasive Screens

Ideally, high-throughput screens should not damage the plants investigated so that they can be used directly for the next steps in forward genetics. Besides visual screens for altered growth or coloration phenotypes, other non-invasive screening procedures have been employed. For instance, systematic measurements of chlorophyll fluorescence have been used to determine the efficiency of photosynthetic electron flow. To this end, image analysis [101, 102] and a semi-automated pulse-amplitude modulated fluorometer device [103] have been developed. Chlorophyll fluorometry has also been used for the quantitative assessment of drought survival [104]. The luciferase and green fluorescent protein (GFP; and its derivatives) reporter genes allow one to monitor changes in the expression of genes by measuring luminescence and fluorescence, respectively, and these have been utilized to identify mutants with deregulated gene expression or altered subcellular localisation of gene products (*e.g.* [105, 106]). Especially the luciferase screen with its very short response time can be employed to test for transient inductions. This system has been applied in stress or ABA-related mutant screens [107, 108], as well as in circadian rhythm related mutant screens [109-111].

Some of the lines generated within the SAIL project were transformed with a vector that carries the LAT52:GUS reporter gene, a visible cell-autonomous marker for pollen grains and tubes [68]. This population was generated in the *qrt1-2* mutant background -- a mutant, which maintains the four male meiotic products in a tetrad [112] and which is very useful for non-invasive screens during male gametophyte development. Mutant lines can be classified as either homozygous or hemizygous for their T-DNA insertion by staining their pollen grains. Consequently, the *hapless* (*hap*) mutation was isolated, which impairs the development or function of haploid gametophytes [113].

The collection of GAL4 enhancer trap lines harbors a GAL4-responsive GFP gene as a marker to tag specific cell types and to reveal developmental transitions [114]. The lines can be screened microscopically and are very useful to identify enhancers for a specific cell type. The screen has been successfully applied and resulted in the isolation of Dof elements, which are involved in the regulation of guard cell gene expression [115]. In a further step some well characterized lines with specific expression patterns can be used as starter lines for EMS mutagenesis. This approach was applied to dissect the mechanisms that specify the different populations of pericycle cells. Thus, a pericycle-specific enhancer trap line was mutagenized and several mutants exhibiting qualitative or quantitative alteration of the GFP expression pattern were isolated [116].

SYSTEMATIC REVERSE GENETICS SCREENS

Efficient systematic reverse genetics requires the availability of gene-indexed mutant populations, and the use of robust, high-throughput assays for phenotypic characterizations. Studies of this type in *Arabidopsis* have benefitted from previous experience with the model organisms *Caenorhabditis elegans* [117] and *Saccharomyces cerevisiae* [118]. In particular, T-DNA mutant collections have been widely used for reverse genetics, as indicated by the more than 2000 cumulative citations in the literature [9]. The majority of genes in the *Arabidopsis* genome are members of gene families [1], so that redundant gene functions often have to be assessed by generating double or higher-order mutants from T-DNA lines.

In the course of systematic reverse genetic screens, lines from different collections (see Table **1**) are usually combined, in order to achieve maximum coverage of the target genes. In addition, for genes that are not covered by existing mutant collections, partial loss-of-function lines are often generated by the knock-down approach (see Section 2.2).

Small- and Medium-Scale Screens

Numerous small- and medium-scale reverse genetic screens have been performed which have addressed specific biological functions or multigene families. Only a small selection can be discussed here (Table **3**). One of the first systematic screens with insertion-tagged populations was performed more than 20 years ago. A collection of 8,000 *Arabidopsis* plants carrying 48,000 insertions of the maize transposable element *En-1* was screened for knock-out alleles of genes involved in flavonoid biosynthesis, utilizing a PCR-based screening protocol [119]. Another early example involved the systematic analysis of nuclear genes for photosystem I subunits, and used T-DNA or transposon insertion lines, together with knock-down lines [120, 121].

For reverse genetic analysis of the nine members of the 1-aminocyclopropane-1-carboxylate synthase (ACS) gene family, seven T-DNA mutants and two amiRNA lines were employed [122] (Table **3**). A larger screen was conducted for subtilisin-like serine proteases (subtilases), which comprises 56 members in *Arabidopsis*. Here 144 T-DNA insertion lines for 55 subtilase genes were identified. With the exception of *SDD1*, none of the lines displayed an obvious visible phenotype in the homozygous state [123]. This result highlights the need to use robust, high-throughput phenotypic screening methods to assign a phenotype, *i.e.* a biological function, to each gene, even in medium-scale screens.

Table 3. Selected reverse genetics screens in *Arabidopsis.*

Gene Family/Biological Process	Numbers of Screened Genes	References
Small scale screen (less than 20 genes)		
NADP-malic enzyme / respiration	4	[126]
Sucrose synthase / carbon metabolism	6	[127]
Type A response regulator / cytokinin signaling	6 of 10	[128]
Aminocyclopropane-1-carboxylate synthase (ACS) gene family / ethylene biosynthesis	9	[122]
Photosystem I subunit (nuclear encoded) / photosynthesis	11	[120, 121]
nucleobase-ascorbate transporter / unknown biological function	12	[129]
Laccase / glyocoprotein	12 of 17	[130]
Auxin response factor / auxin-mediated transcriptional activation/repression	18 of 23	[131]
Medium scale screens (20 to 100 genes)		
chloroplast protein kinases and phosphatases / reversible protein phosphorylation	22	[132-134]
Pentatricopeptide repeat proteins / organelle biogenesis	39 of 450	[135]
Subtilases / serine proteases	55	[123]
pollen tube growth	33	[136]
P450 cytochrome oxidases	91	[137]
Large-scale screens (more than 100 genes)		
Chloroplast function	2733	[138]
Chloroplast function	1369	[139]
Leaf development	Over 700	[140]
Metabolomics	Over 200	[141]

In some cases, where genes with a known overall function were characterised by reverse genetics, an in-depth description could be provided for every member of the set. This was achieved, for instance, for all genes coding for photosystem I subunits (see above; reviewed in: [120]). The opposite case is represented by the so-called "guilt-by-association" approach, which is based on the concept that genes with similar functions often behave similarly with respect to their associated transcript, protein or metabolite profiles as measured by –omics technologies. Co-expression analysis combines gene expression measurements from a large number of experiments thereby providing improved statistics. Thus, co-expression analysis can be used to predict the function of genes based on the similarity of their expression patterns [124, 125].

For instance, it was shown that nuclear genes for chloroplast proteins involved in photosynthesis or plastid gene expression indeed exhibit a high level of co-expression at the transcript level [142]. This finding was then exploited to systematically characterize by reverse genetics genes of unknown function that exhibited photosynthesis gene-like transcriptional profiles, and led to the identification of PGRL1, the elusive ferredoxin-plastoquinone reductase in photosynthetic cyclic elctron flow [143, 144]. A gene for a plastidic bile-acid transporter was also isolated based on a co-expression approach [90]. Moreover, a set of genes, which are specifically expressed in pollen tubes in response to their growth in the pistil but not expressed during other stages of pollen or plant development, was identified. For 33 pollen tube–expressed genes the respective mutants were analyzed in a reverse genetics approach. The mutants were derived from a subset of the SAIL collection that was generated in the *quartet* (*qrt*) background containing a LAT52 promoter driven GUS reporter gene [68]. Indeed, seven out of 33 investigated genes were critical for pollen tube growth [136]. Other examples are reviewed in [145]. However, the "guilt-by-association" approach often involves screening large numbers of genes, and very few mutants exhibit the "desired" phenotype.

An example of chemical genetic screening in combination with reverse genetics involves *Arabidopsis* lines overexpressing 91 P450 cytochrome oxidases. In the course of this screen two P450 cytochrome oxidases that hydroxylate the therapeutic compound 8-methoxypsorelen were identified [137]. This demonstrates the potential of chemical genetic screening in combination with reverse genetics to assign specific functions to individual members of a large protein family.

Large-Scale Screens

In the course of the Chloroplast 2010 Project mutations in nuclear genes that were computationally predicted to encode chloroplast-targeted proteins were subjected to a diverse set of phenotypic screens [138, 146, 147]. During this study a total of 2733 mutants were found to be homozygous for the inspected T-DNA insertion, and 85 phenotypic characters were evaluated in each. The phenotypic tests involved quantitative measurements of metabolites, such as fatty acid methylesters in leaves or amino acid profiles in seeds, and quantification of photosynthetic parameters by a chlorophyll fluorescence assay [146]. Data are available on the project's website (see Table **2**). Among the confirmed mutants obtained by the Chloroplast 2010 Project are mutants for At1g10310 (previously described as a pterin aldehyde reductase involved in folate metabolism) that were found to have an altered fatty acid desaturation phenotype. Single insertions in the gene *Acyl Carrier Protein4* (*ACP4*) were responsible for defects in growth, fatty acid profiles and chlorophyll fluorescence, previously associated with the knockdown of the *Acyl Carrier Protein* gene family. However, the case studies of the Chloroplast 2010 Project also stressed the difficulty of securely establishing genotype-phenotype relationships. It became clear, that -- as in traditional reverse genetic screens -- it is essential to confirm the genotype in advance and to exclude the possibility that a secondary mutation causes the phenotype. A second independent collection of 3246 *Ds*/*Spm*- or T-DNA-tagged homozygous lines for 1369 nuclear-encoded chloroplast proteins has been initiated [139]. Phenotypes were monitored in 3-week-old seedlings grown on agar plates and images are presented on the Chloroplast Function Database (http://rarge.psc.riken.jp/chloroplast/).

Leaf development was addressed using the Salk Unimutant collection [140]. Over 700 mutants with variations in many different leaf traits (*e.g.* form, size, number and pigmentation) were isolated, described and the information deposited in the PhenoLeaf database (http://genetics.umh.es/phenoleaf).

In a different approach, a consortium is profiling the metabolome of specific T-DNA knockout alleles (over 200 mutants to date), detecting about 1800 predetermined metabolites per line and providing the information in a searchable database (http://plantmetabolomics.vrac.iastate.edu/ver2/index.php, [141, 148]). This example shows that also brute-force screens can be adapted to profiling piplines.

NEW TRENDS

The Unimutant Collection – not Quite the Ultimate Tool?

The isolation of two homozygous insert lines for every gene in the *Arabidopsis* genome would make it possible to systematically test the phenotypic consequences of gene loss under a wide variety of conditions by forward genetics [52]. The SALK unimutant collection covers as of April 2015 24,535 individual genes from about 26,000 protein-coding genes, with the aim to identify two homozygote mutants for every gene. Once the unimutant collection is complete, it can be screened on several large-scale platforms for high-throughput assays (*e.g.* [149-152]). The major advantage of screens based on the unimutant collection is obvious: exhaustive screens can be carried out with high efficiency, because loss-of-function lines for all *Arabidopsis* genes can be tested relatively quickly. But loss-of-function lines have their limitations. For instance, they provide only the first step in the analysis of essential genes (see Section 2.2). Secondly, the function of many genes will be very difficult to assess because of their apparent phenotypic silence in many assays. Thirdly, genetic redundancy will continue to hamper attempts to assign functions to members of gene families (see Section 5.2).

From the Unimutant to the Uni-Multimutant Collection: A Closer Approach to the Ultimate Tool

How can the problem of genetic redundancy be solved for thousands of *Arabidopsis* genes? In the simplest case, when two unlinked genes code for the same protein, the systematic generation of double mutants will make such genes manageable. In fact, in the course of the German Plant Genome Initiative (GABI), one ongoing project is dedicated to the systematic generation of segmentally duplicated genes. Segmental duplications account for the duplication of around 2900 genes in the *Arabidopsis* genome and can encompass chromosomal segments with up to about 300 genes which have been duplicated as blocks [153, 154]. In the GABI-DUPLO project, corresponding double mutants have been generated for about 400 such gene pairs by crossing of T-DNA-based single mutants, selfing and PCR-based genotyping, representing a valuable complement to the unimutant collection ([155], https://www.gabi-kat.de/duplo.html).

Closely linked gene duplications, including tandem duplications as well as gene families containing more than two closely related genes, can only be handled by gene silencing approaches, and here the AGRIKOLA collection or amiRNA lines (see Section 2.2) promises to meet this need.

Targeted Mutagenesis

Site-directed mutagenesis remains a significant technical challenge in higher plants. Many attempts have been made to establish targeted mutagenesis in *Arabidopsis*, but no systematic approach is available yet [156].

The first tool developed were zinc-finger nucleases (ZFNs) which were used for targeted gene inactivation in *Arabidopsis* [157]. In this study it was shown that

ZFNs driven by a heat-shock promoter could be integrated into the genome. Upon heat-activation the target gene of the ZFN could be modified at frequencies of around 3%. Subsequently, two studies showed that genes could also be directly targeted by ZFNs [158, 159].

The use of transcription activator-like effectors (TALEs) represents another approach with great potential. TALEs are injected into plant cells *via* the type III secretion pathway found in many plant pathogenic *Xanthomonas* spp. Once inside, they may contribute to disease or trigger resistance by binding to DNA and turning on TALE-specific host genes. Recent advances in understanding of the mechanistic basis of specific DNA binding by TALEs [160] should permit engineering of TALEs for arbitrary gene activation or as translational fusions for directing other proteins, such as negative regulators, methylases or nucleases, to specific DNA targets (reviewed in [161]).

RNA-guided genome editing becomes possible also with the CRISPR-Cas9 system from *Streptococcus pyogenes in Arabidopsis* [162, 163]. A guide RNA directs Cas9 to the genomic sequence that is to be modified. The target is recognized by base-pairing between the guide RNA and the genome, leading to a high specificity. This technique can be used to target genes, for which no other mutants are available, but at the moment it only allows single modifications and no collection is (yet) available.

Cataloguing of Mutant Phenotypes

Because very many research groups generate data from forward and reverse genetics experiments, a central depository for documentation of mutant phenotypes is highly desirable. To this end, The Arabidopsis Information Resource (TAIR) recently began to assign a phenotype to each mutant allele and made pictorial records of the mutant phenotypes available online (Table **2**; [164]). Curators at TAIR have been capturing phenotypic information from the *Arabidopsis* literature since 2005 by adding distinct free-text phenotypes associated with genotypes derived from almost publications.

Phenotypic databases already exist for individual mutant collections. These include, for instance, the Riken *Arabidopsis* Phenome Information Database (RAPID) (Table **2**), which is a searchable database describing morphological phenotypes of ~4,000 *Ds* transposon mutant lines [31]. Phenotypic descriptions of these lines have been classified into eight primary categories (such as seedlings, leaves, stems, flowers and siliques) and fifty secondary categories. Images of individual plants are also available and can be searched by the line number or the phenotypic categories. The PHENOPSIS DB provides phenotypic data from experiments on the PHENOPSIS platform and provides information on how certain genotypes behave in different environmental conditions by providing precise climatic conditions of growth. As of April 2015 more than 57,000 phenotypic data from 17 experiments with 1055 genotypes were available on the website (http://bioweb.supagro.inra.fr/phenopsis/, [165]).

The Future of Systematic Phenotypic Screens

Assuming that one needs to screen a minimum of two homozygous individuals for two different mutant alleles for each of the approximately 28,000 genes in *Arabidopsis*, a total of 112,000 individual plants would have to be planted to analyse the completed unimutant collection (see Section 5.1). When grown in standard trays, a total greenhouse area of 420 m^2 would be needed -- a requirement which becomes feasible if plants are screened consecutively and not in parallel. A space-saving alternative are screens which can be performed on seedlings using either media plates or liquid assays in media-containing multi-well plates, to which different compounds to generate abiotic or biotic stresses can be easily added. Moreover, because the unimutant collection would need to be complemented with double mutants for segmentally duplicated genes (like the GABI-DUPLO lines) and RNAi lines for members of very similar members of multigene families (like AGRIKOLA or amiRNA lines), the numbers listed above might have to be increased by 25% to screen a uni-multimutant collection. Exhaustive characterisation of such populations can only be achieved by the collaborations involving many laboratories, each covering a specific set of phenotypic screens according to their specific expertise. Such screens would in many cases be repetitions of already established phenotypic assays but, because of the large number of plants to be considered, automated, innovative and non-invasive screens will have to be developed. Such screens will combine the capture of 'classical' visible phenotypes (such as growth rate, plant and root architecture and flowering time), specialized non-invasive assays (such as the ones to measure photosynthesis) and 'chemical' phenotyping (such as the profiling of metabolites, proteins and transcripts). In addition, the response of plants to different environmental conditions will have to be assessed. Finally, data collection and processing has to be streamlined to allow efficient and reliable assignment of phenotypes to individual plants. This will have to include the capability of following and documenting the fate (and phenotype) of each individual plant from sowing to disposal, preferably in an automated way. New resources for the computational analysis of images and data sets are important. In addition, standardized storage of data that allows systematic interpretation of the phenotyping results is necessary. The use of ontology terms to describe the phenotypes -- like the use of Gene Ontology to describe biological processes, molecular functions and cellular components -- will have to be employed.

In spite of these many challenges, the prediction that genome-wide phenotypic data will soon become available for *Arabidopsis* does not appear to be all that far-fetched. Its realisation will no doubt stimulate a wealth of in-depth gene-function analyses.

CONFLICT OF INTEREST

The author confirms that he has no conflict of interest to declare for this publication.

ACKNOWLEDGEMENTS

We thank the German Ministry for Education and Research (BMBF) and the German Science Foundation (DFG) for funding (GABI-DUPLO; SFB-TR1 TP8) and Paul Hardy for critical reading of the manuscript.

DISCLOSURE

This is an updated chapter of the earlier review article [by Bolle, C., Schneider, A., & Leister, D. entitled "Perspectives on systematic analyses of gene function in Arabidopsis thaliana: new tools, topics and trends". *Current Genomics*, 2011; 12(1), 1] with additional text and references.

REFERENCES

[1] Arabidopsis Genome Initiative. Analysis of the genome sequence of the flowering plant *Arabidopsis thaliana*. Nature 2000; 408(6814): 796-815.

[2] Saito K, Hirai MY, Yonekura-Sakakibara K. Decoding genes with coexpression networks and metabolomics - 'majority report by precogs'. Trends Plant Sci 2008; 13(1): 36-43.

[3] Allen E, Moing A, Ebbels TM, Maucourt M, Tomos AD, Rolin D, *et al.* Correlation Network Analysis reveals a sequential reorganization of metabolic and transcriptional states during germination and gene-metabolite relationships in developing seedlings of *Arabidopsis*. BMC Syst Biol 2010; 4: 62.

[4] Hannah MA, Caldana C, Steinhauser D, Balbo I, Fernie AR, Willmitzer L. Combined transcript and metabolite profiling of *Arabidopsis* grown under widely variant growth conditions facilitates the identification of novel metabolite-mediated regulation of gene expression. Plant Physiol 2010; 152(4): 2120-9.

[5] Alonso JM, Ecker JR. Moving forward in reverse: genetic technologies to enable genome-wide phenomic screens in *Arabidopsis*. Nat Rev Genet 2006; 7(7): 524-36.

[6] Ostergaard L, Yanofsky MF. Establishing gene function by mutagenesis in *Arabidopsis thaliana*. Plant J 2004; 39(5): 682-96.

[7] Hilson P. Cloned sequence repertoires for small- and large-scale biology. Trends Plant Sci 2006; 11(3): 133-41.

[8] Bouche N, Bouchez D. *Arabidopsis* gene knockout: phenotypes wanted. Curr Opin Plant Biol 2001; 4(2): 111-7.

[9] Wang YH. How effective is T-DNA insertional mutagenesis in *Arabidopsis*? J Biochem Tech 2008; 1(1): 11-20.

[10] Peters JL, Cnudde F, Gerats T. Forward genetics and map-based cloning approaches. Trends Plant Sci 2003; 8(10): 484-91.

[11] Schneeberger K, Ossowski S, Lanz C, Juul T, Petersen AH, Nielsen KL, *et al.* SHOREmap: simultaneous mapping and mutation identification by deep sequencing. Nat Methods 2009; 6(8): 550-1.

[12] Singer T, Fan Y, Chang HS, Zhu T, Hazen SP, Briggs SP. A high-resolution map of *Arabidopsis* recombinant inbred lines by whole-genome exon array hybridization. PLoS Genet 2006; 2(9): e144.

[13] Colbert T, Till BJ, Tompa R, Reynolds S, Steine MN, Yeung AT, *et al.* High-throughput screening for induced point mutations. Plant Physiol 2001; 126(2): 480-4.

[14] Kurowska M, Daszkowska-Golec A, Gruszka D, Marzec M, Szurman M, Szarejko I, *et al.* TILLING: a shortcut in functional genomics. J Appl Genet 2011; 52(4): 371-90.

[15] Oleykowski CA, Bronson Mullins CR, Godwin AK, Yeung AT. Mutation detection using a novel plant endonuclease. Nucleic Acids Res 1998; 26(20): 4597-602.

[16] Till BJ, Colbert T, Tompa R, Enns LC, Codomo CA, Johnson JE, *et al.* High-throughput TILLING for functional genomics. Methods Mol Biol 2003; 236: 205-20.

[17] Tsai H, Missirian V, Ngo KJ, Tran RK, Chan SR, Sundaresan V, *et al.* Production of a high- efficiency TILLING population through polyploidization. Plant Physiol 2013; 161(4): 1604-14.

[18] Lai KS, Kaothien-Nakayama P, Iwano M, Takayama S. A TILLING resource for functional genomics in *Arabidopsis thaliana* accession C24. Genes Genet Syst 2012; 87(5): 291-7.

[19] Martin B, Ramiro M, Martinez-Zapater JM, Alonso-Blanco C. A high-density collection of EMS-induced mutations for TILLING in Landsberg erecta genetic background of *Arabidopsis*. BMC Plant Biol 2009; 9: 147.

[20] Ossowski S, Schneeberger K, Clark RM, Lanz C, Warthmann N, Weigel D. Sequencing of natural strains of *Arabidopsis thaliana* with short reads. Genome Res 2008; 18(12): 2024-33.

[21] Weigel D, Mott R. The 1001 genomes project for *Arabidopsis thaliana*. Genome Biol 2009; 10(5): 107.

[22] Cao J, Schneeberger K, Ossowski S, Gunther T, Bender S, Fitz J, *et al.* Whole-genome sequencing of multiple *Arabidopsis thaliana* populations. Nat Genet 2011; 43(10): 956-63.

[23] Alonso JM, Stepanova AN, Leisse TJ, Kim CJ, Chen H, Shinn P, *et al.* Genome-wide insertional mutagenesis of *Arabidopsis thaliana*. Science 2003; 301(5633): 653-7.

[24] Sessions A, Burke E, Presting G, Aux G, McElver J, Patton D, *et al.* A high-throughput *Arabidopsis* reverse genetics system. Plant Cell 2002; 14(12): 2985-94.

[25] Woody ST, Austin-Phillips S, Amasino RM, Krysan PJ. The WiscDsLox T-DNA collection: an *Arabidopsis* community resource generated by using an improved high-throughput T-DNA sequencing pipeline. J Plant Res 2007; 120(1): 157-65.

[26] Rosso MG, Li Y, Strizhov N, Reiss B, Dekker K, Weisshaar B. An *Arabidopsis thaliana* T-DNA mutagenized population (GABI-Kat) for flanking sequence tag-based reverse genetics. Plant Mol Biol 2003; 53(1-2): 247-59.

[27] Tissier AF, Marillonnet S, Klimyuk V, Patel K, Torres MA, Murphy G, *et al.* Multiple independent defective suppressor-mutator transposon insertions in *Arabidopsis*: a tool for functional genomics. Plant Cell 1999; 11(10): 1841-52.

[28] Sundaresan V, Springer P, Volpe T, Haward S, Jones JD, Dean C, *et al.* Patterns of gene action in plant development revealed by enhancer trap and gene trap transposable elements. Genes Dev 1995; 9(14): 1797-810.

[29] Nishal B, Tantikanjana T, Sundaresan V. An inducible targeted tagging system for localized saturation mutagenesis in *Arabidopsis*. Plant Physiol 2005; 137(1): 3-12.

[30] Martienssen RA. Functional genomics: probing plant gene function and expression with transposons. Proc Natl Acad Sci U S A 1998; 95(5): 2021-6.

[31] Kuromori T, Wada T, Kamiya A, Yuguchi M, Yokouchi T, Imura Y, *et al.* A trial of phenome analysis using 4000 Ds-insertional mutants in gene-coding regions of *Arabidopsis*. Plant J 2006; 47(4): 640-51.

[32] Ito T, Motohashi R, Kuromori T, Mizukado S, Sakurai T, Kanahara H, *et al.* A new resource of locally transposed Dissociation elements for screening gene-knockout lines *in silico* on the *Arabidopsis* genome. Plant Physiol 2002; 129(4): 1695-9.

[33] Ito T, Motohashi R, Kuromori T, Noutoshi Y, Seki M, Kamiya A, *et al.* A resource of 5,814 dissociation transposon-tagged and sequence-indexed lines of *Arabidopsis* transposed from start loci on chromosome 5. Plant Cell Physiol 2005; 46(7): 1149-53.

[34] Kuromori T, Hirayama T, Kiyosue Y, Takabe H, Mizukado S, Sakurai T, *et al.* A collection of 11 800 single-copy Ds transposon insertion lines in *Arabidopsis*. Plant J 2004; 37(6): 897-905.

[35] Sakurai T, Satou M, Akiyama K, Iida K, Seki M, Kuromori T, *et al.* RARGE: a large-scale database of RIKEN *Arabidopsis* resources ranging from transcriptome to phenome. Nucleic Acids Res 2005; 33(Database issue): D647-50.

[36] Samson F, Brunaud V, Balzergue S, Dubreucq B, Lepiniec L, Pelletier G, *et al.* FLAGdb/FST: a database of mapped flanking insertion sites (FSTs) of *Arabidopsis thaliana* T-DNA transformants. Nucleic Acids Res 2002; 30(1): 94-7.

[37] Samson F, Brunaud V, Duchene S, De Oliveira Y, Caboche M, Lecharny A, *et al.* FLAGdb^{++}: a database for the functional analysis of the *Arabidopsis* genome. Nucleic Acids Res 2004; 32(Database issue): D347-50.

[38] Hilson P, Small I, Kuiper MT. European consortia building integrated resources for *Arabidopsis* functional genomics. Curr Opin Plant Biol 2003; 6(5): 426-9.

[39] Ichikawa T, Nakazawa M, Kawashima M, Iizumi H, Kuroda H, Kondou Y, *et al.* The FOX hunting system: an alternative gain-of-function gene hunting technique. Plant J 2006; 48(6): 974-85.

[40] Weigel D, Ahn JH, Blazquez MA, Borevitz JO, Christensen SK, Fankhauser C, *et al.* Activation tagging in *Arabidopsis*. Plant Physiol 2000; 122(4): 1003-13.

[41] Nakazawa M, Ichikawa T, Ishikawa A, Kobayashi H, Tsuhara Y, Kawashima M, *et al.* Activation tagging, a novel tool to dissect the functions of a gene family. Plant J 2003; 34(5): 741-50.

[42] Ichikawa T, Nakazawa M, Kawashima M, Muto S, Gohda K, Suzuki K, *et al.* Sequence database of 1172 T-DNA insertion sites in *Arabidopsis* activation-tagging lines that showed phenotypes in T1 generation. Plant J 2003; 36(3): 421-9.

[43] Robinson SJ, Tang LH, Mooney BA, McKay SJ, Clarke WE, Links MG, *et al.* An archived activation tagged population of *Arabidopsis thaliana* to facilitate forward genetics approaches. BMC Plant Biol 2009; 9: 101.

[44] Schneider A, Kirch T, Gigolashvili T, Mock HP, Sonnewald U, Simon R, *et al.* A transposon- based activation-tagging population in *Arabidopsis thaliana* (TAMARA) and its application in the identification of dominant developmental and metabolic mutations. FEBS Lett 2005; 579(21): 4622-8.

[45] Mitsuda N UY, Ikeda M, Shikata M, Koyama T, Matsui K, Narumi T, Aida R, Sasaki K, Hiyama T, Higuchi Y, Ono M, Isuzugawa K, Saito K, Endo R, Ikeda K, Nakatsuka T, Nishihara M, Yamamura S, Yamamura T, Terakawa T, Ohtsubo N and Ohme-Takagi M. FioreDB: a database of phenotypic information induced by the chimeric repressor silencing technology (CRES-T) in *Arabidopsis* and floricultural plants. *Plant Biotechnology* 2008; 25: 37-43.

[46] Luo C, Lam E. Chromatin charting: global mapping of epigenetic effects. Methods Mol Biol 2009; 553: 127-39.

[47] Lam E, Luo C, Watanabe N. Charting functional and physical properties of chromatin in living cells. Curr Opin Genet Dev 2009; 19(2): 135-41.

[48] Parinov S, Sundaresan V. Functional genomics in *Arabidopsis*: large-scale insertional mutagenesis complements the genome sequencing project. Curr Opin Biotechnol 2000; 11(2): 157-61.

[49] Lepage E, Zampini E, Boyle B, Brisson N. Time- and cost-efficient identification of T-DNA insertion sites through targeted genomic sequencing. PLoS One 2013; 8(8): e70912.

[50] O'Malley RC, Barragan CC, Ecker JR. A User's Guide to the *Arabidopsis* T-DNA Insertion Mutant Collections. Methods Mol Biol 2015; 1284: 323-42.

[51] Ulker B, Peiter E, Dixon DP, Moffat C, Capper R, Bouche N, *et al.* Getting the most out of publicly available T-DNA insertion lines. Plant J 2008; 56(4): 665-77.

[52] O'Malley RC, Ecker JR. Linking genotype to phenotype using the *Arabidopsis* unimutant collection. Plant J 2010; 61(6): 928-40.

[53] Ossowski S, Schwab R, Weigel D. Gene silencing in plants using artificial microRNAs and other small RNAs. Plant J 2008; 53(4): 674-90.

[54] Green PJ, Pines O, Inouye M. The role of antisense RNA in gene regulation. Annu Rev Biochem 1986; 55: 569-97.

[55] Chuang CF, Meyerowitz EM. Specific and heritable genetic interference by double-stranded RNA in *Arabidopsis thaliana*. Proc Natl Acad Sci U S A 2000; 97(9): 4985-90.

[56] Hilson P, Allemeersch J, Altmann T, Aubourg S, Avon A, Beynon J, *et al.* Versatile gene-specific sequence tags for *Arabidopsis* functional genomics: transcript profiling and reverse genetics applications. Genome Res 2004; 14(10B): 2176-89.

[57] Schwab R, Ossowski S, Riester M, Warthmann N, Weigel D. Highly specific gene silencing by artificial microRNAs in *Arabidopsis*. Plant Cell 2006; 18(5): 1121-33.

[58] Schwab R, Ossowski S, Warthmann N, Weigel D. Directed gene silencing with artificial microRNAs. Methods Mol Biol 2010; 592: 71-88.

[59] Hiratsu K, Matsui K, Koyama T, Ohme-Takagi M. Dominant repression of target genes by chimeric repressors that include the EAR motif, a repression domain, in *Arabidopsis*. Plant J 2003; 34(5): 733-9.

[60] Ito T, Meyerowitz EM. Overexpression of a gene encoding a cytochrome P450, CYP78A9, induces large and seedless fruit in *arabidopsis*. Plant Cell 2000; 12(9): 1541-50.

[61] Nakazawa M, Yabe N, Ichikawa T, Yamamoto YY, Yoshizumi T, Hasunuma K, *et al.* *DFL1*, an auxin-responsive *GH3* gene homologue, negatively regulates shoot cell elongation and lateral root formation, and positively regulates the light response of hypocotyl length. Plant J 2001; 25(2): 213-21.

[62] Seki M, Shinozaki K. Functional genomics using RIKEN *Arabidopsis thaliana* full-length cDNAs. J Plant Res 2009; 122(4): 355-66.

[63] Tani H, Chen X, Nurmberg P, Grant JJ, SantaMaria M, Chini A, *et al.* Activation tagging in plants: a tool for gene discovery. Funct Integr Genomics 2004; 4(4): 258-66.

[64] Marsch-Martinez N, Greco R, Van Arkel G, Herrera-Estrella L, Pereira A. Activation tagging using the En-I maize transposon system in *Arabidopsis*. Plant Physiol 2002; 129(4): 1544-56.

[65] Wilson K, Long D, Swinburne J, Coupland G. A Dissociation insertion causes a semidominant mutation that increases expression of *TINY*, an *Arabidopsis* gene related to *APETALA2*. Plant Cell 1996; 8(4): 659-71.

[66] Pogorelko GV, Fursova OV, Ogarkova OA, Tarasov VA. A new technique for activation tagging in *Arabidopsis*. Gene 2008; 414(1-2): 67-75.

[67] Rosin FM, Watanabe N, Cacas JL, Kato N, Arroyo JM, Fang Y, *et al.* Genome-wide transposon tagging reveals location-dependent effects on transcription and chromatin organization in *Arabidopsis*. Plant J 2008; 55(3): 514-25.

[68] McElver J, Tzafrir I, Aux G, Rogers R, Ashby C, Smith K, *et al.* Insertional mutagenesis of genes required for seed development in *Arabidopsis thaliana*. Genetics 2001; 159(4): 1751-63.

[69] Tzafrir I, Dickerman A, Brazhnik O, Nguyen Q, McElver J, Frye C, *et al.* The *Arabidopsis* SeedGenes Project. Nucleic Acids Res 2003; 31(1): 90-3.

[70] Tzafrir I, Pena-Muralla R, Dickerman A, Berg M, Rogers R, Hutchens S, *et al.* Identification of genes required for embryo development in *Arabidopsis*. Plant Physiol 2004; 135(3): 1206-20.

[71] Meinke D, Muralla R, Sweeney C, Dickerman A. Identifying essential genes in *Arabidopsis thaliana*. Trends Plant Sci 2008; 13(9): 483-91.

[72] Budziszewski GJ, Lewis SP, Glover LW, Reineke J, Jones G, Ziemnik LS, *et al.* *Arabidopsis* genes essential for seedling viability: isolation of insertional mutants and molecular cloning. Genetics 2001; 159(4): 1765-78.

[73] Dobritsa AA, Geanconteri A, Shrestha J, Carlson A, Kooyers N, Coerper D, *et al.* A large-scale genetic screen in *Arabidopsis* to identify genes involved in pollen exine production. Plant Physiol 2011; 157(2): 947-70.

[74] De Muyt A, Pereira L, Vezon D, Chelysheva L, Gendrot G, Chambon A, *et al.* A high throughput genetic screen identifies new early meiotic recombination functions in *Arabidopsis thaliana*. PLoS Genet 2009; 5(9): e1000654.

[75] Yang WC, Sundaresan V. Genetics of gametophyte biogenesis in *Arabidopsis*. Curr Opin Plant Biol 2000; 3(1): 53-7.

[76] Drews GN, Yadegari R. Development and function of the angiosperm female gametophyte. Annu Rev Genet 2002; 36: 99-124.

[77] Johnson MA, Preuss D. Plotting a course: multiple signals guide pollen tubes to their targets. Dev Cell 2002; 2(3): 273-81.

[78] Wilson ZA, Morroll SM, Dawson J, Swarup R, Tighe PJ. The *Arabidopsis MALE STERILITY1 (MS1)* gene is a transcriptional regulator of male gametogenesis, with homology to the PHD- finger family of transcription factors. Plant J 2001; 28(1): 27-39.

[79] Sorensen AM, Krober S, Unte US, Huijser P, Dekker K, Saedler H. The *Arabidopsis ABORTED MICROSPORES (AMS)* gene encodes a MYC class transcription factor. Plant J 2003; 33(2): 413- 23.

[80] Feldmann KA, Coury DA, Christianson ML. Exceptional segregation of a selectable marker (KanR) in *Arabidopsis* identifies genes important for gametophytic growth and development. Genetics 1997; 147(3): 1411-22.

[81] Bonhomme S, Horlow C, Vezon D, de Laissardiere S, Guyon A, Ferault M, *et al.* T-DNA mediated disruption of essential gametophytic genes in *Arabidopsis* is unexpectedly rare and cannot be inferred from segregation distortion alone. Mol Gen Genet 1998; 260(5): 444-52.

[82] Christensen CA, Subramanian S, Drews GN. Identification of gametophytic mutations affecting female gametophyte development in *Arabidopsis*. Dev Biol 1998; 202(1): 136-51.

[83] Howden R, Park SK, Moore JM, Orme J, Grossniklaus U, Twell D. Selection of T-DNA-tagged male and female gametophytic mutants by segregation distortion in *Arabidopsis*. Genetics 1998; 149(2): 621-31.

[84] Procissi A, de Laissardiere S, Ferault M, Vezon D, Pelletier G, Bonhomme S. Five gametophytic mutations affecting pollen development and pollen tube growth in *Arabidopsis thaliana*. Genetics 2001; 158(4): 1773-83.

[85] Huck N, Moore JM, Federer M, Grossniklaus U. The *Arabidopsis* mutant *feronia* disrupts the female gametophytic control of pollen tube reception. Development 2003; 130(10): 2149-59.

[86] Lalanne E, Honys D, Johnson A, Borner GH, Lilley KS, Dupree P, *et al.* SETH1 and SETH2, two components of the glycosylphosphatidylinositol anchor biosynthetic pathway, are required for pollen germination and tube growth in *Arabidopsis*. Plant Cell 2004; 16(1): 229-40.

[87] Pagnussat GC, Yu HJ, Ngo QA, Rajani S, Mayalagu S, Johnson CS, *et al.* Genetic and molecular identification of genes required for female gametophyte development and function in *Arabidopsis*. Development 2005; 132(3): 603-14.

[88] Boavida LC, Shuai B, Yu HJ, Pagnussat GC, Sundaresan V, McCormick S. A collection of Ds insertional mutants associated with defects in male gametophyte development and function in *Arabidopsis thaliana*. Genetics 2009; 181(4): 1369-85.

[89] Portemer V, Renne C, Guillebaux A, Mercier R. Large genetic screens for gynogenesis and androgenesis haploid inducers in *Arabidopsis thaliana* failed to identify mutants. Front Plant Sci 2015; 6: 147.

[90] Gigolashvili T, Yatusevich R, Rollwitz I, Humphry M, Gershenzon J, Flügge UI. The plastidic bile acid transporter 5 is required for the biosynthesis of methionine-derived glucosinolates in *Arabidopsis thaliana*. Plant Cell 2009; 21(6): 1813-29.

[91] Fujita M, Mizukado S, Fujita Y, Ichikawa T, Nakazawa M, Seki M, *et al.* Identification of stress-tolerance-related transcription-factor genes via mini-scale Full-length cDNA Over-eXpressor (FOX) gene hunting system. Biochem Biophys Res Commun 2007; 364(2): 250-7.

[92] Papdi C, Abraham E, Joseph MP, Popescu C, Koncz C, Szabados L. Functional identification of *Arabidopsis* stress regulatory genes using the controlled cDNA overexpression system. Plant Physiol 2008; 147(2): 528-42.

[93] Papdi C, Leung J, Joseph MP, Salamo IP, Szabados L. Genetic screens to identify plant stress genes. Methods Mol Biol 2010; 639: 121-39.

[94] Jing B, Xu S, Xu M, Li Y, Li S, Ding J, *et al.* Brush and spray: a high-throughput systemic acquired resistance assay suitable for large-scale genetic screening. Plant Physiol 2011; 157(3): 73- 80.

[95] McCourt P, Desveaux D. Plant chemical genetics. New Phytol 2010; 185(1): 15-26.

[96] Tsuchiya Y, Vidaurre D, Toh S, Hanada A, Nambara E, Kamiya Y, *et al.* A small-molecule screen identifies new functions for the plant hormone strigolactone. Nat Chem Biol 2010; 6(10): 741-9.

[97] Park SY, Fung P, Nishimura N, Jensen DR, Fujii H, Zhao Y, *et al.* Abscisic acid inhibits type 2C protein phosphatases via the PYR/PYL family of START proteins. Science 2009; 324(5930): 1068-71.

[98] Li J, Mu J, Bai J, Fu F, Zou T, An F, *et al.* PARAQUAT RESISTANT1, a Golgi-localized putative transporter protein, is involved in intracellular transport of paraquat. Plant Physiol 2013; 162(1): 470-83.

[99] Zhao Y, Chow TF, Puckrin RS, Alfred SE, Korir AK, Larive CK, *et al.* Chemical genetic interrogation of natural variation uncovers a molecule that is glycoactivated. Nat Chem Biol 2007; 3(11): 716-21.

[100] Girke T, Cheng LC, Raikhel N. ChemMine. A compound mining database for chemical genomics. Plant Physiol 2005; 138(2): 573-7.

[101] Niyogi KK, Grossman AR, Bjorkman O. *Arabidopsis* mutants define a central role for the xanthophyll cycle in the regulation of photosynthetic energy conversion. Plant Cell 1998; 10(7): 1121-34.

[102] Shikanai T, Munekage Y, Shimizu K, Endo T, Hashimoto T. Identification and characterization of *Arabidopsis* mutants with reduced quenching of chlorophyll fluorescence. Plant Cell Physiol 1999; 40(11): 1134-42.

[103] Varotto C, Pesaresi P, Maiwald D, Kurth J, Salamini F, Leister D. Identification of photosynthetic mutants of *Arabidopsis* by automatic screening for altered effective quantum yield of photosystem 2. PHOTOSYNTHETICA 2000; 38(4): 497-504.

[104] Woo NS, Badger MR, Pogson BJ. A rapid, non-invasive procedure for quantitative assessment of drought survival using chlorophyll fluorescence. Plant Methods 2008; 4: 27.

[105] Baruah A, Simkova K, Hincha DK, Apel K, Laloi C. Modulation of O-mediated retrograde signaling by the PLEIOTROPIC RESPONSE LOCUS 1 (PRL1) protein, a central integrator of stress and energy signaling. Plant J 2009; 60(1): 22-32.

[106] Tanaka H, Kitakura S, De Rycke R, De Groodt R, Friml J. Fluorescence imaging-based screen identifies ARF GEF component of early endosomal trafficking. Curr Biol 2009; 19(5): 391-7.

[107] Chinnusamy V, Stevenson B, Lee BH, Zhu JK. Screening for gene regulation mutants by bioluminescence imaging. Sci STKE 2002; 2002(140): pl10.

[108] Foster R, Chua NH. An *Arabidopsis* mutant with deregulated ABA gene exprcssion: implications for negative regulator function. Plant J 1999; 17(4): 363-72.

[109] Southern MM, Millar AJ. Circadian genetics in the model higher plant, *Arabidopsis thaliana*. Methods Enzymol 2005; 393: 23-35.

[110] Kevei E, Gyula P, Hall A, Kozma-Bognar L, Kim WY, Eriksson ME, *et al.* Forward genetic analysis of the circadian clock separates the multiple functions of ZEITLUPE. Plant Physiol 2006; 140(3): 933-45.

[111] Onai K, Okamoto K, Nishimoto H, Morioka C, Hirano M, Kami-Ike N, *et al.* Large-scale screening of *Arabidopsis* circadian clock mutants by a high-throughput real-time bioluminescence monitoring system. Plant J 2004; 40(1): 1-11.

[112] Preuss D, Rhee SY, Davis RW. Tetrad analysis possible in *Arabidopsis* with mutation of the QUARTET (QRT) genes. Science 1994; 264(5164): 1458-60.

[113] Johnson MA, von Besser K, Zhou Q, Smith E, Aux G, Patton D, *et al. Arabidopsis hapless* mutations define essential gametophytic functions. Genetics 2004; 168(2): 971-82.

[114] Haseloff J, Dormand EL, Brand AH. Live imaging with green fluorescent protein. Methods Mol Biol 1999; 122: 241-59.

[115] Gardner MJ, Baker AJ, Assie JM, Poethig RS, Haseloff JP, Webb AA. GAL4 GFP enhancer trap lines for analysis of stomatal guard cell development and gene expression. J Exp Bot 2009; 60(1): 213-26.

[116] Parizot B, Laplaze L, Ricaud L, Boucheron-Dubuisson E, Bayle V, Bonke M, *et al.* Diarch symmetry of the vascular bundle in *Arabidopsis* root encompasses the pericycle and is reflected in distich lateral root initiation. Plant Physiol 2008; 146(1): 140-8.

[117] Piano F, Schetter AJ, Mangone M, Stein L, Kemphues KJ. RNAi analysis of genes expressed in the ovary of Caenorhabditis elegans. Curr Biol 2000; 10(24): 1619-22.

[118] Giaever G, Chu AM, Ni L, Connelly C, Riles L, Veronneau S, *et al.* Functional profiling of the Saccharomyces cerevisiae genome. Nature 2002; 418(6896): 387-91.

[119] Wisman E, Hartmann U, Sagasser M, Baumann E, Palme K, Hahlbrock K, *et al.* Knock-out mutants from an En-1 mutagenized *Arabidopsis thaliana* population generate phenylpropanoid biosynthesis phenotypes. Proc Natl Acad Sci U S A 1998; 95(21): 12432-7.

[120] Jensen PE, Bassi R, Boekema EJ, Dekker JP, Jansson S, Leister D, *et al.* Structure, function and regulation of plant photosystem I. Biochim Biophys Acta 2007; 1767(5): 335-52.

[121] Pesaresi P, Varotto C, Richly E, Kurth J, Salamini F, Leister D. Functional genomics of *Arabidopsis* photosynthesis. Plant Physiol Biochem 2001; 39: 285-94.

[122] Tsuchisaka A, Yu G, Jin H, Alonso JM, Ecker JR, Zhang X, *et al.* A combinatorial interplay among the 1-aminocyclopropane-1-carboxylate isoforms regulates ethylene biosynthesis in *Arabidopsis thaliana*. Genetics 2009; 183(3): 979-1003.

[123] Rautengarten C, Steinhauser D, Bussis D, Stintzi A, Schaller A, Kopka J, *et al.* Inferring hypotheses on functional relationships of genes: Analysis of the *Arabidopsis thaliana* subtilase gene family. PLoS Comput Biol 2005; 1(4): e40.

[124] Usadel B, Obayashi T, Mutwil M, Giorgi FM, Bassel GW, Tanimoto M, *et al.* Co-expression tools for plant biology: opportunities for hypothesis generation and caveats. Plant Cell Environ 2009; 32(12): 1633-51.

[125] Loraine A. Co-expression analysis of metabolic pathways in plants. Methods Mol Biol 2009; 553: 247-64.

[126] Wheeler MC, Tronconi MA, Drincovich MF, Andreo CS, Flügge UI, Maurino VG. A comprehensive analysis of the NADP-malic enzyme gene family of *Arabidopsis*. Plant Physiol 2005; 139(1): 39-51.

[127] Bieniawska Z, Paul Barratt DH, Garlick AP, Thole V, Kruger NJ, Martin C, *et al.* Analysis of the sucrose synthase gene family in *Arabidopsis*. Plant J 2007; 49(5): 810-28.

[128] To JP, Haberer G, Ferreira FJ, Deruere J, Mason MG, Schaller GE, *et al.* Type-A *Arabidopsis* response regulators are partially redundant negative regulators of cytokinin signaling. Plant Cell 2004; 16(3): 658-71.

[129] Maurino VG, Grube E, Zielinski J, Schild A, Fischer K, Flügge UI. Identification and expression analysis of twelve members of the nucleobase-ascorbate transporter (NAT) gene family in *Arabidopsis thaliana*. Plant Cell Physiol 2006; 47(10): 1381-93.

[130] Cai X, Davis EJ, Ballif J, Liang M, Bushman E, Haroldsen V, *et al.* Mutant identification and characterization of the laccase gene family in *Arabidopsis*. J Exp Bot 2006; 57(11): 2563-9.

[131] Okushima Y, Overvoorde PJ, Arima K, Alonso JM, Chan A, Chang C, *et al.* Functional genomic analysis of the *AUXIN RESPONSE FACTOR* gene family members in *Arabidopsis thaliana*: unique and overlapping functions of *ARF7* and *ARF19*. Plant Cell 2005; 17(2): 444-63.

[132] Schliebner I, Pribil M, Zühlke J, Dietzmann A, Leister D. A Survey of Chloroplast Protein Kinases and Phosphatases in *Arabidopsis thaliana*. Curr Genomics 2008; 9(3): 184-90.

[133] Bonardi V, Pesaresi P, Becker T, Schleiff E, Wagner R, Pfannschmidt T, *et al.* Photosystem II core phosphorylation and photosynthetic acclimation require two different protein kinases. Nature 2005; 437(7062): 1179-82.

[134] Pribil M, Pesaresi P, Hertle A, Barbato R, Leister D. Role of plastid protein phosphatase TAP38 in LHCII dephosphorylation and thylakoid electron flow. PLoS Biol 2010; 8(1): e1000288.

[135] Lurin C, Andres C, Aubourg S, Bellaoui M, Bitton F, Bruyere C, *et al.* Genome-wide analysis of *Arabidopsis* pentatricopeptide repeat proteins reveals their essential role in organelle biogenesis. Plant Cell 2004; 16(8): 2089-103.

[136] Qin Y, Leydon AR, Manziello A, Pandey R, Mount D, Denic S, *et al.* Penetration of the stigma and style elicits a novel transcriptome in pollen tubes, pointing to genes critical for growth in a pistil. PLoS Genet 2009; 5(8): e1000621.

[137] Kruse T, Ho K, Yoo HD, Johnson T, Hippely M, Park JH, *et al.* In planta biocatalysis screen of P450s identifies 8-methoxypsoralen as a substrate for the CYP82C subfamily, yielding original chemical structures. Chem Biol 2008; 15(2): 149-56.

[138] Ajjawi I, Lu Y, Savage LJ, Bell SM, Last RL. Large-scale reverse genetics in *Arabidopsis*: case studies from the Chloroplast 2010 Project. Plant Physiol 2010; 152(2): 529-40.

[139] Myouga F, Akiyama K, Motohashi R, Kuromori T, Ito T, Iizumi H, *et al.* The Chloroplast Function Database: a large-scale collection of *Arabidopsis* Ds/Spm- or T-DNA-tagged homozygous lines for nuclear-encoded chloroplast proteins, and their systematic phenotype analysis. Plant J 2010; 61(3): 529-42.

[140] Wilson-Sanchez D, Rubio-Diaz S, Munoz-Viana R, Perez-Perez JM, Jover-Gil S, Ponce MR, *et al.* Leaf phenomics: a systematic reverse genetic screen for *Arabidopsis* leaf mutants. Plant J 2014; 79(5): 878-91.

[141] Bais P, Moon-Quanbeck SM, Nikolau BJ, Dickerson JA. Plantmetabolomics.org: mass spectrometry-based *Arabidopsis* metabolomics--database and tools update. Nucleic Acids Res 2012; 40(Database issue): D1216-20.

[142] Biehl A, Richly E, Noutsos C, Salamini F, Leister D. Analysis of 101 nuclear transcriptomes reveals 23 distinct regulons and their relationship to metabolism, chromosomal gene distribution and co-ordination of nuclear and plastid gene expression. Gene 2005; 344: 33-41.

[143] DalCorso G, Pesaresi P, Masiero S, Aseeva E, Schünemann D, Finazzi G, *et al.* A complex containing PGRL1 and PGR5 is involved in the switch between linear and cyclic electron flow in *Arabidopsis*. Cell 2008; 132(2): 273-85.

[144] Hertle AP, Blunder T, Wunder T, Pesaresi P, Pribil M, Armbruster U, *et al.* PGRL1 is the elusive ferredoxin-plastoquinone reductase in photosynthetic cyclic electron flow. Mol Cell 2013; 49(3): 511-23.

[145] Armbruster U, Pesaresi P, Pribil M, Hertle A, Leister D. Update on chloroplast research: new tools, new topics and new trends. Mol Plant 2010; 4(1):1-16.
[146] Lu Y, Savage LJ, Ajjawi I, Imre KM, Yoder DW, Benning C, *et al.* New connections across pathways and cellular processes: industrialized mutant screening reveals novel associations between diverse phenotypes in *Arabidopsis*. Plant Physiol 2008; 146(4): 1482-500.
[147] Lu Y, Savage LJ, Larson MD, Wilkerson CG, Last RL. Chloroplast 2010: a database for large- scale phenotypic screening of *Arabidopsis* mutants. Plant Physiol 2011; 155(4): 1589-600.
[148] Bais P, Moon SM, He K, Leitao R, Dreher K, Walk T, *et al.* PlantMetabolomics.org: a web portal for plant metabolomics experiments. Plant Physiol 2010; 152(4): 1807-16.
[149] Kuromori T, Takahashi S, Kondou Y, Shinozaki K, Matsui M. Phenome analysis in plant species using loss-of-function and gain-of-function mutants. Plant Cell Physiol 2009; 50(7): 1215-31.
[150] Finkel E. Imaging. With 'phenomics,' plant scientists hope to shift breeding into overdrive. Science 2009; 325(5939): 380-1.
[151] Yazdanbakhsh N, Fisahn J. Analysis of *Arabidopsis thaliana* root growth kinetics with high temporal and spatial resolution. Ann Bot 2010; 105(5): 783-91.
[152] Joosen RV, Kodde J, Willems LA, Ligterink W, van der Plas LH, Hilhorst HW. GERMINATOR: a software package for high-throughput scoring and curve fitting of *Arabidopsis* seed germination. Plant J 2010; 62(1): 148-59.
[153] Vision TJ, Brown DG, Tanksley SD. The origins of genomic duplications in *Arabidopsis*. Science 2000; 290(5499): 2114-7.
[154] Blanc G, Hokamp K, Wolfe KH. A recent polyploidy superimposed on older large-scale duplications in the *Arabidopsis* genome. Genome Res 2003; 13(2): 137-44.
[155] Bolle C, Huep G, Kleinbölting N, Haberer G, Mayer K, Leister D, *et al.* GABI-DUPLO: a collection of double mutants to overcome genetic redundancy in *Arabidopsis thaliana*. Plant J 2013; 75(1): 157-71.
[156] Puchta H, Fauser F. Gene targeting in plants: 25 years later. Int J Dev Biol 2013; 57(6-8): 629- 37.
[157] Osakabe K, Osakabe Y, Toki S. Site-directed mutagenesis in *Arabidopsis* using custom- designed zinc finger nucleases. Proc Natl Acad Sci U S A 2010; 107(26): 12034-9.
[158] de Pater S, Pinas JE, Hooykaas PJ, van der Zaal BJ. ZFN-mediated gene targeting of the *Arabidopsis* protoporphyrinogen oxidase gene through Agrobacterium-mediated floral dip transformation. Plant Biotechnol J 2013; 11(4): 510-5.
[159] Qi Y, Li X, Zhang Y, Starker CG, Baltes NJ, Zhang F, *et al.* Targeted deletion and inversion of tandemly arrayed genes in *Arabidopsis thaliana* using zinc finger nucleases. G3 (Bethesda) 2013; 3(10): 1707-15.
[160] Boch J, Bonas U, Lahaye T. TAL effectors--pathogen strategies and plant resistance engineering. New Phytol 2014; 204(4): 823-32.
[161] Bogdanove AJ, Schornack S, Lahaye T. TAL effectors: finding plant genes for disease and defense. Curr Opin Plant Biol 2010; 13(4): 394-401.
[162] Feng Z, Zhang B, Ding W, Liu X, Yang DL, Wei P, *et al.* Efficient genome editing in plants using a CRISPR/Cas system. Cell Res 2013; 23(10): 1229-32.
[163] Li JF, Zhang D, Sheen J. Targeted Plant Genome Editing via the CRISPR/Cas9 Technology. Methods Mol Biol 2015; 1284: 239-55.
[164] Swarbreck D, Wilks C, Lamesch P, Berardini TZ, Garcia-Hernandez M, Foerster H, *et al.* The *Arabidopsis* Information Resource (TAIR): gene structure and function annotation. Nucleic Acids Res 2008; 36(Database issue): D1009-14.
[165] Fabre J, Dauzat M, Negre V, Wuyts N, Tireau A, Gennari E, *et al.* PHENOPSIS DB: an information system for *Arabidopsis thaliana* phenotypic data in an environmental context. BMC Plant Biol 2011; 11: 77.

CHAPTER 8

Genetic Engineering for Cold Stress Tolerance in Crop Plants

Shabir H. Wani[1,*], Saroj Kumar Sah[2], Gulzar Sanghera[4], Waseem Hussain[5] and N.B. Singh[3]

[1]Division of Genetics and Plant Breeding, SKUAST-K, Srinagar-190025, Jammu and Kashmir, India; [2]Department of Biochemistry, Molecular Biology, Entomology and Plant Pathology, Mississippi State University, MS-39762, USA; [3]Department of, Plant Breeding and Genetics, COA, CAU, Imphal, 795001, Manipur India; [4]RRS, Punjab Agricultural University, Post Box No. 34 Kapurthala, Punjab, 144601, India and [5]Department of Agronomy & Horticulture. University of Nebraska, Lincoln, 68583-0915, USA

Abstract: Plants respond to low temperatures with changes in their pattern of gene expression and protein products. Many species of tropical or subtropical origin are injured or killed by non-freezing low temperature, and exhibit various symptoms of chilling injury such as chlorosis, necrosis, or growth retardation. In contrast, chilling tolerant species are able to grow at such cold temperatures. Conventional breeding methods have met with limited success in improving the cold tolerance of important crop plants through inter-specific or inter-generic hybridization. Recent studies involving full genome profiling/sequencing, mutational and transgenic plant analyses have provided a deep insight of the complex transcriptional mechanism that operates under cold stress. The alterations in expression of genes in response to cold temperatures are followed by increases in the levels of hundreds of metabolites, some of which are known to have protective effects against the damaging effects of cold stress. Various low temperature inducible genes have been isolated from plants. Most appear to be involved in tolerance to cold stress and the expression of some of them is regulated by C-repeat binding factor/dehydration-responsive element binding (*CBF/DREB*1) transcription factors. Numerous physiological and molecular changes occur during cold acclimation reveals that the cold resistance is more complex than perceived and involves more than one pathway. The findings summarized in this review have shown potential practical applications for breeding cold tolerance in crop and horticultural plants suitable to temperate geographical locations.

Keywords: Biosynthesis, cold acclimatization, cold stress, crop plants, Fructan, genes, genetic engineering, glycine-betaine, Growth hormone, Hill agriculture, low temperature, morpho-physiological, pathways, proline, RNA chaperon, stress inducible promoter, transcription factors, Transporter protein.

***Corresponding author Shabir Hussain Wani:** Division of Genetics and Plant Breeding, SKUAST-K, Srinagar-190025, Jammu and Kashmir, India; Tel:+91 9419035566; E-mail: shabirhussainwani@gmail.com

Atta-ur-Rahman (Ed.)

1. INTRODUCTION

A series of morphological, physiological, biochemical and molecular changes triggers in plants due to the adverse effect of abiotic stresses on growth and productivity of plants. Expression of a variety of genes is induced by different stresses in diverse plants. Cold stress is a major environmental factor that limits the agricultural productivity of plants in hilly areas. Plants respond and adapt to this stress conditions through changes at the molecular,cellular, physiological and biochemical levels.

Low temperature stress causes significant crop losses by affecting plant growth and crop productivity [1, 2]. Plants differ in their tolerance to chilling (0-15 °C) and freezing (<0°C) temperatures. In general, plants from temperate climatic regions are considered to be chilling tolerant with variable degree, and can increase their freezing tolerance by being exposed to chilling, non-freezing temperatures, a process known as cold acclimation [3], which is associated with bio- chemical and physiological changes [4-6] and ultimately showed marked changes in gene expression, biomembrane lipid composition, and small molecule accumulation [7].

On the other hand, plants of tropical and subtropical origins are sensitive to chilling stress and lack the mechanism of cold acclimation. Low temperature resistance in plants is a very complex trait, involving many different metabolic pathways and cell compartments [8]. Improving the cold tolerance of important crop plants through inter-specific or inter-generic hybridization in conventional breeding have met with limited success. Besides, *in vitro* induced variations have also been applied to improve the abiotic stress tolerance of various crop plants but without much success. The conventional breeding approaches are limited by the complexity of stress tolerance traits, low genetic variance of yield components under stress condition and lack of efficient selection criteria. It is important, therefore, to look for alternative strategies to develop cold stress tolerant crops.

Biotechnology offers strategies to develop crop plants with improved tolerance to cold stress through genetic engineering. Rapid advance in recombinant DNA technology and development of precise and efficient gene transfer protocols have resulted in efficient transformation and generation of transgenic lines in a number of crop species [9-11] (Fig. **1**). Transgenic approach is being pursued actively throughout the world to improve traits including tolerance to biotic and abiotic stresses in a number of crops [15]. Many studies have suggested that the genes which express during cold stress are critical for both chilling tolerance [12] and cold acclimation [13, 14]. Advent of molecular tools has made it possible to select directly at the gene label without waiting for the phenotype to show up. As compared to other stresses, plant responses to cold stress are complex, so the prospects of improving cold tolerance in crops seem not to be very bright. Despite this, efforts have been made during the last two decades to generate transgenic lines of different crops, which have shown improved tolerance to cold stress. Therefore, it is important to use most appropriate tools that help in reaching the goals. The improved genotype should be better than the available ones and must reach the farmers. An attempt has been made in this article to review the various mechanisms and genes involved in cold acclimatization and the possibilities where transgenic technology has been explored for breeding cold tolerance in crop plants.

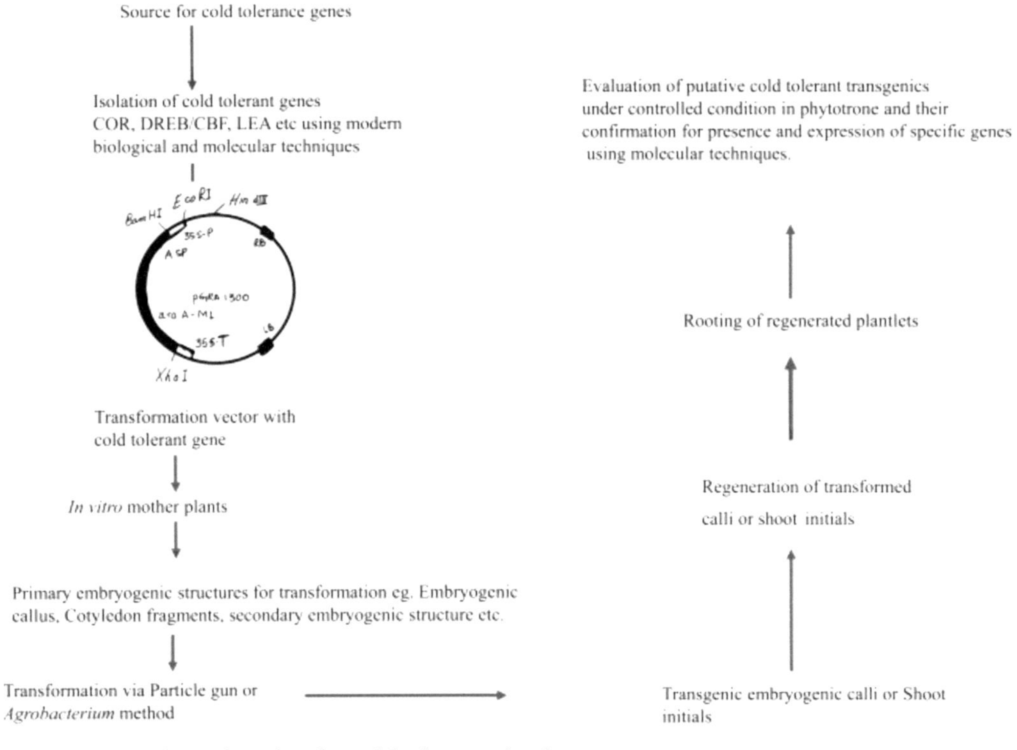

Figure 1: Genetic engineering for cold tolerance in plants.

2. MORPHO-PHYSIOLOGICAL BASIS OF COLD TOLERANCE

A large number of studies have evaluated different plant species tolerant to different stresses such as drought, salinity and cold. However, less detail is given with regard to the methods used to evaluate the stress response; these studies may bring about some misleading conclusions from an agronomic or physiology perspective [16]. In order to closely mimic the changes induced through stress, plants should be given cycles of stress, rather than short exposure to very severe stresses; although if the purpose is to assess gene expression only short exposures to stress are certainly adequate. In this section, we focus on the agronomic/physiological perspective. We are not challenging the quality of the work done to assess gene expression. Our intention is to try to reconcile both approaches (agronomic and molecular) toward a common focus: breeding cold tolerance.

The temperate and cool regions are those where altitudes ranged from 1600-2500 m amsl (above mean sea level) and temperature during crop growth period ranged from 5-20 °C [17]. There is wide range of cold stress in temperate areas differing in both timing and intensity of low temperature. The primary abiotic stress which limits the crop productivity in temperate regions is low temperature. It is a major problem at seedling and reproductive stages, results in slow establishment and low seed set which leads to poor yield of the crop [16]. The low temperature limits the crop productivity when temperatures remain above freezing that is > 0 °C, it is called as chilling stress. Chilling sensitive cultivars are typically tropical genotypes. Yield losses are more severe when cold stress occurs during reproductive stage/ anthesis in rice which lead to high spikelet sterility [18]. Ability of crop genotypes / lines to survive /perform better under low temperature than other genotypes is called as cold tolerance. Ordinarily, it is the consequence of cold hardening that is an earlier exposure to a low temperature for a specific period as a result of which chilling tolerance of the concerned plants increases. Cold tolerance involves increased chlorophyll accumulation, reduced sensitivity of photosynthesis, improved germination, pollen fertility and seed set which are desirable as:

2.1. Increased Chlorophyll Accumulation

Chlorophyll accumulation inhibits in actively growing leaves under low temperature. For example, cold tolerant *japonica* lines in rice accumulates more chlorophyll under cold stress than do cold sensitive lines of *indica* rice [19]. Rasolofo [20] evaluated 181 rice accessions to identify donor and outstanding cold tolerant lines using leaf discolouration score (1-3) and found 19 remained green (dark) after 10 days in the 12 °C cold water tank. Sanghera *et al.* [18] found 18 cold tolerance IRCTN rice genotypes based on dark green colour and high spikelet fertility (>90%) under temperate conditions. Liu *et al.* [21] conducted the experiment under low temperature of 1°C and −10 °C and found that chlorophyll content of leaves of naked oats was decreased during prolonged low temperature stress and they reported that naked oats was more damage at −10 ^{o}C as compared to 1 ^{o}C. Similarly, Janmohammadi *et al.* [22] reported that there is a positive correlation between chlorophyll content and low temperature tolerance from an experiment conducted on winter wheat and ray.

2.2. Reduced Sensitivity of Photosynthesis

Chloroplast and photosynthesis is major site of cold injury. Tolerance in these aspects is expressed in native vegetation adapted to growing under cool conditions. The reduced sensitivity of photosynthesis to cold has been observed in maize inbreds adapted to low temperature which is partly related to specific enzymes of the process [23]. Calvin cyclase is the enzyme that plays major role to maintain photosynthetic activity in cold-acclimated plants [24]. This is in agreement with Janmohammadi *et al.* [25], who reported that many changes occurs during cold acclimation were related to photosynthetic enzymes by proteome analysis of hardy winter wheat whereas their expression pattern was same in autumn as well as in winter.

2.3. Improved Germination

Genetic variation in cold tolerance at germination and seedling stage has been documented. Saini and Tandon [26] found that L62G, Heng Jodo, Jodo, Heugdo, IR AT 102, Khonorullo, K 78-28, Daegaldo, Mujudo and L62-2A genotypes were cold tolerant having more than 85% germination and good seedling vigour (score 3) at an average of 11 ^{o}C field temperature.

2.4. Improved Pollen Fertility and Seed Set

Cold tolerance at reproductive stage is expressed as improved seed set and pollen fertility. Lia *et al.* [27] reported that cold tolerance in rice is associated with anther size, number of pollen grains and diameter of fertile pollen grains at booting stage. However, Sanghera *et al.* [18] reported that cold tolerance is associated with high spikelet fertility (>90%) and well panicle exsertion under temperate conditions.

Cold stress triggers a reaction in the plant that prevents starch accumulation in the pollen, it leads to the failure of pollen germination due to non aviability of energy provided by starch. Eventually, the pollination will not occur and hence the grain will not produce. CSIRO has found that all the ingredients for starch are present but they are not getting into the pollen grain where they are needed. A cell layer surrounding the pollen, called the 'tapetum', is responsible for feeding the pollen with sugar. The tapetum get active only for 1-2 days so if a cold snap occurs during this time, the pollen will not grow further. The sugar cannot freely move into the tapetum and pass through it to the pollen. Instead the sugar is brokendown in bits with the help of a catalyst known as 'Invertase' and then transported to the pollen through tapetum [28]. Quantities of invertase are decreased in conventional rice and remain at normal levels in a cold tolerant variety when both are exposed to cold temperatures. By comparing a cold tolerant strain of rice with conventional rice, CSIRO has found that the gene responsible for invertase seems exactly the same in both strains. So it was concluded that the invertase gene itself does not make the rice plant cold tolerant – but instead a mechanism that regulates the invertase gene is different. Early research indicates that the invertase gene is regulated by the hormone abscisic acid (ABA). Oliver *et al.* [29] reported that the rice plants behaved like they have experienced a cold snap and became sterile when they were injected with ABA in the absence of any cold stress. Also, ABA levels increase when conventional rice is exposed to cold, but they remain the same in the cold tolerant variety. Recent studies have indicated that the difference between cold-sensitive and tolerant rice is due to a different ability to control ABA levels [30]. It has also been shown that this mechanism may require interactions with other plant hormones like auxins [31]. Further, Zhao *et al.* [30] also reported that low temperature turns off the genes responsible for sugar transport into the pollen grains and therefore starch cannot be produced in the pollen in cold conditions. The sugar metabolism genes continued to function normally during cold treatment of cold tolerant rice, hence the cold tolerant varieties like chinese rice produced fertile pollen following cold treatment. Ample genetic variation for cold tolerance is available in well adapted breeding population. Germplasm collected from high altitude and low temperature areas, cold tolerant mutant, somaclonal variants and wild species can be exploited for breeding improved cold tolerant genotypes in hilly areas [16].

3. MECHANISMS FOR UNDERSTANDING TOLERANCE TO COLD INJURY

There is a huge impact of low temperature on the survival and geographical distribution of plants. Depending on the intensity and duration of the stress a range of cellular metabolisms in plant system will get affected. When exposed to low temperatures, plants respond with changes in their pattern of gene expression and protein products. Different studies have indicated that the primary site of freezing injury in plants are the membrane systems of the cells. [3, 32]. In addition, the severe dehydration due to freezing is the primary cause of freeze-induced membrane damage [32, 33]. Many species of tropical or subtropical origin are injured or killed by nonfreezing low temperatures, and exhibit various symptoms of chilling injury such as chlorosis, necrosis, or growth retardation. In contrast, chilling- tolerant species are able to grow at such cold temperatures.

However, multiple forms of membrane damage like expansion-induced-lysis, lamellar-to-hexagonal-II phase transitions, and fracture jump lesions can occur due to freeze induced cellular dehydration [33]. Thus, the key function of cold acclimation should be to stabilize membranes against freezing injury. Indeed, cold acclimation prevents expansion-induced-lyses and the formation of hexagonal II phase lipids in plants [33]. This ability to adapt under stress has a great impact on crop yields, survival and distribution of the plant. Multiple mechanisms appear to be involved in this stabilization, mostly refers to the changes in lipid composition [33]. Temperature stress also induce change in membrane fluidity in plants and might represent a potential site of perception and/or injury [34, 35]. Cold acclimation function by altering the composition of membrane lipid with increased fatty acid unsaturation. Chloroplast glycerol-3-phosphate acyltransferase (GPAT) gene involved in the desaturation of phosphatidyl glycerol fatty acid was overexpressed in genetically engineered tobacco. The increase in the number of unsaturated fatty acids lead to decrease in the chilling sensitivity. Thus the unsaturation of membrane lipids appers to be crucial for optimum membrane function at low temperature. An *Arabidopsis* fatty acid biosynthesis (*FAB1*) mutant with more saturated membranes showed decreased quantum efficiency of photosystem II (PSII), chlorophyll content and the amount of chloroplast glycerolipids after prolonged exposure to low temperature [36]. A triple mutant for fatty acid desaturation (*fad3-2 fad7-2 fad8*) produced a phenotype similar to *FAB1* in the absence of trienoic fatty acids (18:3 or 16:3), when plants were subjected to prolonged low temperature exposure [37]. Similarly, Arabidopsis f*ad5* and *fad6* mutants with more saturated membranes became chlorotic and showed growth retardation during low temperature incubation [38]. In addition to membrane unsaturation, lipid asymmetry and accumulation of sucrose and other simple sugars in the membrane that typically occurs with cold acclimation also contributes to membrane physical structure at low temperature and protect it from freeze- induced damage *in vitro* [39-41].

In addition, chaperon proteins like late embryogenesis abundant (LEA) proteins act against cellular damage which indicate their role in anti aggregation of enzymes under desiccation and freezing stresses. Certain novel hydrophilic and LEA polypeptides have also been found to participate in the stabilization of membranes against freeze-induced injury. These hydrophilic and LEA polypeptides are predicted to form amphipathic a-helices which helps the intrinsic curvature of monolayers to form hexagonal II phase [42]. In another study it has been reported that freeze-induced production of reactive oxygen species contributes to membrane damage. In suchconditions, the increase in antioxidants level [45], increased levels of sugars in the apoplastic space [46], and the induction of genes encoding molecular chaperones [44] could have protective effects. Besides,adhesion of intercellular ice with cell walls and protein denaturation that occurs in plants at low temperature could potentially result in cellular damage and cell rupture [43, 44]. In cold stress-tolerant plants, many genes involved in the synthesis of osmoprotectants—organic compounds such as amino acids (*e.g.* proline), quaternary and other amines (*e.g.* glycinebetaine and polyamines) and a variety of sugars and sugar alcohols (*e.g.* mannitol, trehalose and galactinol) are induced to protect the cells from injury Thus it is clear that transferring a single gene encoding a single specific stress protein may not be sufficient to reach the required tolerance levels. To enhance tolerance towards multiple stresses a single gene encoding a stress inducible transcription factor that regulates a number of other genes can be used. Nucleic-acid- binding cold shock proteins (CSPs) accumulate in bacteria at cold temperatures and function as transcription antiterminators or translational enhancers by destabilizing RNA secondary structure [47]. Some CSP-domain-containing proteins in plants are upregulated by cold stress, and might function as RNA chaperones in the regulation of translation [48, 49]. Genetic analysis of another cold-responsive nucleic-acid-binding protein from *Arabidopsis* designated as *atRZ-1a* which got upregulated by cold stress, supports its function in freezing tolerance [50]. Helicases have also found to play an important role in cold tolerance in plants. Plants have the largest number of DEAD-box RNA helicase genes as compared to other organisms [51]. One of these helicases, *LOS4* encoded by the *Arabidopsis* osmotically responsive genes, is found to be essential for plant tolerance of chilling and freezing stress [52]. *LOS4* is required for efficient export of RNA from the nucleus to the cytoplasm [53]. In addition, the *Arabidopsis* nucleoporin *AtNUP160,* suppressor of auxin resistance1 (*SAR1*) also controls RNA export, and is crucial for chilling and freezing tolerance [54]. Mutant *LOS4* and *AtNUP160* plants do not have severe growth or developmental phenotypes due to defects in the nucleocytoplasmic transport of RNA which affect cold tolerance preferentially.

4. KEY PLAYERS INVOLVED IN COLD RESPONSIVE PATHWAYS

Cold tolerance is a complex physiological mechanisms which involves many cell and plant traits. Mechanism of cold tolerance in plants is still not fully clear because the genetic control of cold tolerance is complex and can be regarded as polygenic [55] Therefore, the understanding of cold responsive pathways at molecular and genetic level will pave the way for the development of improved methodologies for cold tolerance screening.The discovery of change in gene expression during cold acclimation was the starting of exploration of antifreezing molecular mechanisms. In this context, Zhao *et al.* [30] identified 242 unique genes that are expressed differentially between cold sensitive and cold tolerant rice through gene expression profiling. These genes are involved in processes such as senescence, cell death, male sterility and plant hormone response. Similarly, global transcript profiling of *Arabidopsis* genome indicated that > 10% of genes were regulated during cold acclimation [57-60]. Transcriptome analysis using microarray technology is a powerful for discovering many stress-inducible genes involved in stress response and tolerance [30, 60-62].

In past, it has been reported that genes induced during stress conditions also triggers a signal transduction network in addition to the production of important metabolic proteins for protecting cells from stress [57]. Several stress induced cor genes such as rd29A, cor15A, kin1 and cor6.6 are triggered in response to cold treatment, ABA and water deficit stress in the early stages of the osmotic stress response. Similarly, a cis-acting element, dehydration responsive element (DRE) has identified in many dehydration responsive genes like rd29A and found to be involved in ABA-independent gene expression under drought, low temperature and high salt stress conditions. Thus, the overexpression of some drought-responsive transcription factors can lead to the expression of downstream genes and the enhancement of abiotic stress tolerance in plants. The protein product of these genes are classified into two groups [58, 59]. The first group of proteins that probably function in stress tolerance includes chaperones, LEA proteins, osmotin, antifreeze proteins, some key enzymes for osmolyte biosynthesis (like proline, water channel proteins, sugar and proline transport- ers, detoxification enzymes), mRNA-binding proteins, enzymes for fatty acid metabolism (proteinase inhibitors, ferritin) and lipid-transfer proteins [63]. The second group contains protein factors involved in regulation of signal transduction and gene ex- pression that probably function during stress response [64]. It has been reported that some of these stress- inducible genes (such as genes for enzymes for osmolyte biosynthesis, LEA proteins and detoxification enzymes) have showed stress-tolerant phenotypes when overexpressed in transgenic plants [56, 61]. The transcription factors included under second group constitute gene networks by regulating different stress-inducible genes collectively or separately. Seki *et al.* [64] reported that some of these regulatory pathways are also involved in drought-, cold-, or high-salinity stress responses. Though, the exactfunctions of most of these genes are not fully understood. Thus a precise functional analysis of these stress- inducible transcription factors will provide precise information on the complex regulatory gene networks involved in responses different stresses [30, 65]. Some of these stress-inducible regulatory genes that encode proteins (transcription factors) have been overexpressed in transgenic plants and generate stress- tolerant phenotypes in them [66, 67].

5. GENETIC ENGINEERING FOR INDUCING COLD TOLERANCE

A number of genes are turned on when a plant is subjected to abiotic stress. It leads to increase in levels of several metabolites and proteins which are involved in certain degree of protection to these stresses. The introduction of gene or genes that regulate a specific trait using transgenic technology has opened up many exciting possibilities to improve cold stress in plants [75]. It also improves the genetic potential of plants with the development of specific crop varieties that are more resistant to biotic and abiotic stresses with enhanced nutritional level. A key to progress towards breeding better crops under stress has been to understand the changes in cellular, biochemical and molecular machinery that occur in response to stress which in turn provides new tools and strategies to improve the environmental stress tolerance of crops. Since freezing tolerance is a multigenic trait [68], transformation of a single functional gene appears to have a limited effect on crop freezing tolerance [69]. Because many aspects of cold adaptation process are under transcriptional control, many tran- scription regulatory factors were chosen, hence, genetic engineering for introgression of such genes that are known to be involved in stress response and putative tolerance, might prove to be a faster track towards improving crop varieties for enhanced cold tolerance.

Application of genetic engineering in different crops using genes encoding cold-tolerant proteins and metabolites is found to be an alternate to overcome the abiotic stress induced by low temperature. Chilling tolerant transgenic tobacco (*Nicotiana tabacum*) was developed by over expression of a gene encoding chloroplast *w*3 fatty acid desaturase. The transgenic tobacco plants showed chilling tolerance at 1 ºC for 7 d [70]. In another study,chilling tolerance at 1 ºC for 11 d was observed in transgenic tobacco plants by transferring a gene encoding a non-specific cyano- bacterial desaturase which increase the level of unsaturated fatty acid content in membrane lipids [71]. It has also been reported that alteration of the unsaturation of fatty acids by the over expression of glycerol-3- phosphate acyl transferase also conferred chilling tolerance in transgenic rice plants [72-74]. Thus it was concluded that modifications in lipid composition lead to cold tolerance by stabilizing the cell membranes and preventing cellular leakage.

In most of these studies, either an enzyme or a single gene encoding protective protein was overexpressed under the control of the constitutive 35S cauliflower mosaic virus (*CaMV*) promoter, but there it has although several genes have been shown to function in environmental stress tolerance and response [61]. During last two decades, a number of transcription factor(s) related to cold stress tolerance have been identification and isolated. Such a good number of genes identified in different studies (Table **1**) has made it difficult to decide exactly which genes are more tolerant to cold/ freezing. For transformation purpose the transgene(s) should encode enzymes which are involved in biosynthesis of various osmoprotectants, modification of membrane lipids, LEA protein, and detoxification. The transgene(s) encoding protein factors which induce stress-inducible genes involved in cold stress tolerance and signal transduction for stress responses may be useful for improving the cold tolerance of plants.

Table 1. Cold stress-tolerant transgenic crops.

Gene (s) / Gene Product	Cellular Role	Transgenic Host-Plant	Performance of Transgenic Plants	References
gpat Glycerol 3-phosphate acyltransferase	Fatty acid unsaturation	*N. tabacum*	Transformants showed less chilling damage to photosynthetic activity than the wild type	[91]
sod Superoxide dismutase	Dismutation of toxic reactive oxygen intermediate	*N. tabacum*	Transformants showed 20% higher photosynthetic activity during chilling compared to untransformed plants	[121]
sacB Levan sucrase	Fructan biosynthesis	*N. tabacum*	Transformants were more tolerant to freezing and *PEG*-mediated water stress than the wild type	[122]
cor15a Cold regulated gene	Promotes freezing tolerance	*A. thaliana*	Transformants showed *in vivo* enhanced freezing tolerance of protoplasts and the chloroplasts	[69]
mn-sod Mn-Superoxide	Dismutation of reactiveoxygen	*M. sativa*	Transformants showed reduced injury from water deficit stress and increased winter survival	[123]

Table 1: contd...

dismutase	inter mediates in mitochondria			
gst/gpx Glutathione-Stransferase and glutathione peroxidase	Detoxification of herbicides and toxic substances	*N. tabacum*	Transformants over-expressing *GST/GPX* showed stimulated seedling growth under chilling and salt stress	[124]
cbf1 CRT/DRE binding factor	Transcription factor	*A. thaliana*	Transformants showed regulation of several *cor* genes at the same time and showed freezing tolerance	[81]
dreb1 and dreb2 DRE-binding Protein	Transcription factor	*A. thaliana*	Transformants revealed freezing and dehydration tolerance but caused dwarfed phenotypes in transgenic plants	[86]
WCS120/COR39 CCGAC sequences like CRT/DREs in its promoter	Low temperature regu- lated gene	*Trtricum sativum*	cold inducible in monocotyledonous and dicotyledonous plants	[125]
codA Choline oxidase A	Glycinebetaine biosynthesis	*O. sativa*	Transformants accumulated high levels of glycinebetaine and showed increased tolerance to salt and low temperature stress	[126]
codA Choline oxidase A	Glycinebetaine biosynthesis	*A. thaliana*	Transformants were tolerant to salt and cold	[127]
DREB1A (CBF3) DRE-binding protein	Transcription factor	*Arabidopsis*	Increased salt, drought and cold tolerance in nonacclimated plants	[80]
prodh Proline dehydrogenase	Proline biosynthesis	*A. thaliana*	The antisense transgenics were more tolerant to freezing and high salinity than wild types	[128]
CBF3 DRE-binding protein	Transcription factor	*Arabidopsis*	Increased freezing tolerance of cold- acclimated plants	[6]
ala1 Aminophospholipid ATPase 1	P-type ATPase (Transporter protein)	*A. thaliana*	Transformants showing down regulation results in cold-affected plants that are much smaller than the wild type	[39]
SCOF1 cold-inducible zinc finger protein	Regulator of *SGBF-1* as a transcription factor	*Glycine max*	activate *COR* gene expression and increase freezing tolerance in non-acclimated transgenic plants	[107]
abi3 Abscisic acid induced protein	Transcription factor	*A. thaliana*	Marked increase in expression of low tem- perature-induced freezing tolerance accom- panied by up-regulation of *RAB18, LTI129, LTI130 and LTI178*	[14]

Table 1: contd...

CuCOR19 citrus dehydrin	Inhibition of lipid peroxidation	*N. tabacum*	Increased the cold tolerance	[94]
CBF1/ DREB1b DRE-binding protein	Transcription factor	*O. sativa*	The cold-responsive genes lip5, lip9, and OsDhn1 were up-regulated in the transgenic plants	[129]
DREB1A (rd29A) DRE-binding protein	Stress-inducible pro- moter	*N. tabacum*	Improved drought and low-temperature stress tolerance	[88]
OSISAP1 Zinc-finger protein	Transcription factor	*N. tabacum*	The transcript level of *OSISAP*1 was in- creased to a very high level during a 12-h cold treatment	[130]
Osmyb4	Transcription factor	*Arabidopsis*	Increases chilling and freezing tolerance	[131]
HOS10 Encodes an R2R3-type protein	Transcription factor	*O. sativa*	Enhanced cold tolerance	[132]
ZAT12 C2H2 zinc finger	Transcription factor	*Arabidopsis*	Improved cold acclimation	[60]
Cor15am Chloroplast stromal protein	Stress-inducible pro- moter	*Arabidopsis*	Enhanced cryoprotective activity	[133]
OsMYB3R-2 DNA-binding domain	Transcription factor	*Arabidopsis*	Overexpression of *OsMYB3R-2* leads to increased tolerance to freezing, drought, and salt stress	[98]
ACBP6 Acyl-CoA-binding protein	Decline in phosphati- dylcholine and eleva- tion of phosphatidic acid	*Arabidopsis*	Overexpression of *ACBP6* enhanches freezing tolerance	[134]
OsMYB3R-2 DNA-binding domain	Transcription factor	*O. sativa*	Overexpression of OsMYB3R-2 exhibited enhanced cold tolerance	[135]
AtCSP3 *Cold shock protein*	RNA chaperon	*Arabidopsis*	Transgenic plants conferred enhanced freez- ing tolerance as compared to wild type plants hence demonstrating essential role of RNA chaperones for cold adaptation in higher plants	[136]
MYBS3 DNA-binding repeat MYB	Transcription factor	*O. sativa*	Plays a critical role in cold adaptation in rice	[104]
mybc1 Regulate osmotic stress tolerance	Transcription factor	*Arabidopsis*	Exhibited an increased tolerance to freezing stress	[137]

Table 1: contd…

ThpI Thermal hysteresis proteins (Anti freeze protein)	Transcription factor	*Arabidopsis*	Enhanced low temperature tolerance in transgenic plants was observed by changes of electrolyte leakage activity, malonyldialdehyde and proline contents	[138]
CBF1 CRT/DRE binding factor 1	Transcription factor	*Solanum Lycopersicum*	Detection of higher activity of superoxide dismutase (SOD), higher non-photochemical quenching (NPQ), and lower malondialdehyde (MDA) content in transgenic tomato leaves suggest that CBF1 protein plays an important role in protection of PSII and PSI during low temperature stress at low irradiance	[113]
HVA1	Proline accumulation	*Barley*	Transformants shows cold tolerance in mulberry	[139]
Nt Osm *CBF1, P5CS, APX*	Free proline and Ascorbate	*Nicotiana Tobaccum*	Chilling tolerance	[140]
SLICE 1	Accumulation of antioxidant	*Solanum lycopersicum*	Chilling tolerance and enhanced expression of cold-responsive genes	[141,142]
Zm DREB1	Transcription Factor	*Maize*	Histone modification and Chormatin conformational changes gives chilling stress tolerance	[143]
SoBADH *Betaine Aldehyde Dehydrogenase*	Glycine betaine	*Spinacia oleracea*	Improved tolerance to various abiotic stress like salt, oxidative stress and low temperature	[144]
Arabidopsis C-repeat binding factor 3 (AtCBF3), cold-regulated 15A (AtCOR15A)	Binding factor	*Arabidopsis Thaliana*	Promote cold adaptation process to protect egg plant from chilling stress	[145]
PeAPY2 *Apyrase*	Hydrolysed purine nucleotide	*Arabidopsis*	PeAPY2-transgenic plants maintained plasma membrane integrity and showed reduced cold elicited electrolyte leakage compared to wild types.	[146]

Among the various genes involved in low temperature adaptation and signal transduction, expression of some of them is regulated by C-repeat binding factor/dehydration-responsive element binding (*CBF/ DREB*1) transcription factors. The genes encoding CBF transcription factor represent one of the most significant discoveries in this field. All important crops and few vegetables species have contained this gene [12]. Hajela *et al.* [84] reported that a set of genes that encode a related family of cold-regulated (*COR*) proteins, were used by different researchers to identify a family of *Arabidopsis* transcription factors known as either C-repeat binding factors (*CBF*) (*CBF1, CBF2 and CBF3*) or dehydration responsive element-binding factors (*DREB*) (*DREB1B, DREB1C* and *DREB1A*). Three *CBF/DREB*1 genes (*CBF3/DREB1a*, *CBF1/DREB1b*, and *CBF2/DREB1c*) belonging to the *AP2/DREBP* family of DNA-binding proteins have been identified in *Arabidopsis* [78-80]. Overexpression of *CBF1/DREB1b* and *CBF3/DREB1a* enhances cold tolerance by inducing *COR* (cold regulated) genes and it also causes many biochemical changes, such as the accumulation of sugar and proline [6, 80, 81, 86]. Constitutive overexpressing of a cold inducible transcription factor (*CBF*1; *CRT/DRE* binding protein) in transgenic *Arabidopsis* plants showed tolerance to freezing without any negative effect on the development and growth characteristics [81]. The overexpression of *Arabidopsis CBF*1 (*CRT/DRE* binding protein) has been shown to activate cor homologous genes at non-acclimating temperatures [82]. Transgenic tomato (*Solanum lycopersicum*) containing the *CBF*1 cDNA under the control of a CaMV35S promoter showed improved tolerance to chilling, drought and salt stress but ithad negative effects like dwarf phenotype, reduction in fruit set and seed number per fruit in transgenic plants [12]. In another study, on Arabidopsis (*A. thaliana*) and rice CBF/DREB1-dependent cold response pathway has been shown to play a predominant role in freezing tolerance through the process of cold acclimation. These transcription factors binds to promotor *cis* element *CRT/DRE* and activate the expression of these cold responsive genes [76]. Thus, the *CBF/DREB1* genes are thought to be activators that integrate several components of the cold acclimation response by which plants increase their tolerance to low temperatures after exposure to nonfreezing conditions [6].

In another study it has been reported that the expression of cold shock proteins (CSPs) from bacteria like CspA from *Escherichia coli* and CspB from *Bacillus subtilis*, promotes stress adaptation in multiple plant species [83]. Transgenic rice plants containing transgene encoding CspA and CspB showed improved stress tolerance for a number of abiotic stresses such as cold, heat, and water deficits.

There have been numerous efforts in enhancing tolerance towards multiple stresses such as cold, drought and salt stress in crops other than the model plants like *Arabidopsis*, tobacco and alfalfa. An increased tolerance to freezing and drought in *Arabidopsis* was achieved by overexpressing *CBF*4, a close *CBF*/ *DREB*1 homolog whose expression is rapidly induced during drought stress and by ABA treatment, but not by cold [87]. Subsequently, the overexpression of *DREB*1A has been shown to improve the drought- and low- temperature stress tolerance in tobacco, wheat and groundnut [88, 89]. The use of stress inducible *rd*29A promoter mini- mized the negative effects on plant growth in these crop spe- cies. Further, Gilmour *et al.* [85] have shown that ectopic transgenic overexpression of *CBF1/DREB1B*, *CBF2/ DREB1C* or *CBF3/DREB1A* in *Arabidopsis* activates a suite of *CBF/DREB* target genes at warm temperatures and results in increased freezing and drought tolerance. *CBF* transcripts begin accumulating within 15 minutes of plants being ex- posed to low temperature strongly suggests that the low tem- perature "Thermometer" and "Signal Transducer" are pre- sent at warm non-inducing temperatures. Two cDNA clones that encode DRE-binding proteins, *DREB1A/CBF3* and *DREB2A*, have been isolated by using yeast one-hybrid screening. cDNA clones encoding two *DREB1A* homologs (named *DREB1B/CBF1* and *DREB1C/CBF2*) and one *DREB2A* homolog (*DREB2B*) were also isolated and Ex- pression of the *DREB1A* gene and its two homologs was induced by low-temperature stress, whereas expression of the *DREB2A* gene and its homolog was induced by dehydra- tion. These results indicate that two independent families of DREB proteins, *DREB1* and *DREB2*, function as trans- acting factors in two separate signal transduction pathways under low-temperature and dehydration conditions, respec- tively.

However, Agarwal *et al.* [90] showed that *CBF*s are negatively regulated by an upstream transcription factor, *MYB15* (an *R2R3-MYB* family protein) in *Arabidopsis*. *MYB15* is expressed even in the absence of cold stress, and *MYB15* can bind to *MYB* recognition elements (*MYBRS*) in the promoters of *CBFs*. *MYB15* mutant plants show en- hanced expression of CBFs during cold acclimation and en- hanced freezing tolerance, whereas, transgenic *Arabidopsis* overexpressing *MYB15* showed a decreased expression of *CBFs* and a reduction in freezing tolerance. Thus, *MYB15* is an upstream transcription factor that negatively regulates the expression of *CBFs*.

Transcriptome analysis of *ZAT12*-overexpressing *Arabi- dopsis* revealed that the *ZAT12* regulon consists of at least 24 cold standard set (*COS*) genes, of which nine are cold- induced and 15 are cold-repressed genes [60]. Constitutive overexpression of *Arabidopsis DREB1A* improved drought and low-temperature stress tolerance in tobacco, and regula- tion of transgene expression *via* the stress-inducible *RD29A* promoter minimized the negative effects on plant growth [88].

Similarly, the *Arabidopsis DREB1A* gene was placed under control of the *RD29A* promoter and transferred *via* biolistic transformation into bread wheat [89]. However, constitutive overexpression of the *CBF* genes using the cau- liflower mosaic virus 35S promoter can result in undesirable agronomic traits. In *Arabidopsis*, *CBF* overexpression can cause a ''stunted'' growth phenotype, a decrease in seed yield and a delay in flowering [6, 86].

In the last decade, extensive research efforts have been undertaken to identify and characterize cold-responsive (*COR*) genes and a number of homologous components of the *Arabidopsis CBF* cold response pathway in many plants have been found [7]. Many of these putative orthologs have been structured, analyzed and functionally tested. The ex- pression patterns of the *CBFs* and *CORs* in response to low temperature are similar in a variety of plants species, involv- ing rapid cold-induced expression of the *CBF*s followed by expression of *CBF*-targeted genes that increase freezing tol- erance. Moreover, constitutive overexpression of the *Arabi- dopsis CBF* genes in other plants resulted in increased freez- ing tolerance that have been successfully used to engineer cold stress tolerance in several crop species [7].

Transgenic attempts with other structural genes have also been made with fair degree of success. Genetically engi- neered tobacco plants over-expressing chloroplast glycerol-3-phosphate acyltransferase (*G PAT*) gene (involved in phos-phatidyl glycerol fatty acid desaturation) from squash (*Cu- curbita maxima*) and *A. thaliana* [91] showed an increase in the number of unsaturated fatty acids present in the plant cell wall, which enhance the cold tolerance to the plants during cold stress. Expression of a plant phosphatase (*At PP2CA*) in transgenic *A. thaliana* can accelerate the development of cold acclimation and increase freezing tolerance. It has also shown that transgenic plants expressing a constitutively ac-tive kinase *NPK*1, is more tolerant to chilling and other abi- otic stresses [92]. Pennycooke *et al.* [93] reported that down- regulating a-Gal (a -Galactosidase) in transgenic petunia resulted in an increase in freezing tolerance suggesting that engineering raffinose metabolism by transformation with a - Gal provides an additional method for improving the freez- ing tolerance of plants. The overexpression of genes encod- ing *LEA* proteins can improve the stress tolerance of trans- genic plants. Expression of the citrus gene encoding a *LEA*protein, *CuCOR19* increased the cold tolerance of transgenic tobacco [94]. Likewise, the freezing tolerance of *Arabidopsis* was increased by the ectopic expression of the wheat gene *WCS19* [52], the *Arabidopsis* gene *COR15A* [69], and the co-expression of the genes *RAB18* and *COR47* and *XERO2* and *ERD10* [95]. The freezing tolerance of strawberry leaves was enhanced by expression of the wheat dehydrin gene *WCOR410* [77]. On the other hand, the expression of two cold-induced *LEA* proteins from spinach (Kaye *et al*., 1998) [96] and three desiccation-induced *LEA* proteins from *C. plantagineum* [97] in tobacco did not induce any significant changes in the freezing or drought tolerance of the respective transgenic plants. This may indicate either that not all *LEA* proteins make a significant contribution to plant stress toler- ance, or that they need a particular background to function in, as suggested for transgenic strawberry plants [77]. Kim *et al.* [49] reported that, engineered tobacco with ring zinc finger protein (*RDCPt*) from hot pepper (*Capsicum annuum*) and reported that expression of this gene resulted in improved cold toler- ance in transgenic plants as compared to wild type. Dai *et al.* [98] reported that overexpression of *OsMYB3R*-2 in transgenic *Arabidopsis* increased tolerance to freezing when ex- posed to -8 °C for 10 h. They found that survival after 6 d at normal conditions was 26.8% for the wild-type and 84.5% for transgenic lines. Phenotypically, most transgenic seed- lings were green and could regrow as compared with the wild type; whereas most wild-type seedlings became white and did not regrow after removed to normal conditions. The survival percentage under different low temperatures also showed dramatic difference between the transgenic plants and wild-type plants.

According to Pramanik and Imai [99], *TPP* (trehalose-6- phosphate phosphatase) genes expressed in rice and their expression is induced by cold. Trehalose accumulates rapidly and transiently, which follows the transient induction of *TPP* activity, in rice tissues during chilling stress [99]. Overex- pression of *TPS* (*trehalose-6-phosphate synthase*) and *TPP* genes enhanced the accumulation of trehalose and tolerance to cold stress in transgenic tobacco and rice [100-103]. How- ever, the regulatory mechanism of *TPPs* by cold or other stresses is unclear. In another study, Su *et al.* [104] observed that *MYBS*3 plays a critical role in cold adaptation in rice and necessary for enhancing cold tolerance. They reported that transgenic rice constitutively overexpressing *MYBS*3 toler- ated 4 °C for at least 1 week and exhibited no yield penalty in normal field conditions.

Based on previous studies it has been established that the *CBF* cold responsive pathway is an integral component of the cold acclimation response [7, 68, 105]. However, the transcriptome data showed that additional cold-regulatory pathways also exist [57, 58, 106]. Transcriptome comparisons indicated that only 12% of the cold-responsive genes are certain members of the *CBF* regulon. Moreover, at least 28% of the cold-responsive genes were not regulated by the *CBF* transcription factors, including 15 encoding known or putative transcription factors, indicating that these cold- responsive genes are members of different low-temperature regulons [60].

When overexpressed in *Arabidopsis* and tobacco, the soybean gene *SCOF-1* (encodes a zinc-finger protein) can activate *COR* gene expression and increase freezing toler- ance in non-acclimated transgenic plants, although the *SCOF-1* protein does not directly bind to either the *DRE/CRT* or the ABRE elements [102]. *SCOF-1* interacts with another G-box binding *bZIP* protein, *SGBF-1*. *SGBF-1* can activate ABRE-driven reporter gene expression in *Arabidopsis* leaf protoplasts. Thus, *SCOF-1* may regulate the activity of SGBF-1 as a transcription factor in inducing *COR* gene expression [107].

Forward genetic analysis in *Arabidopsis* identified two transcription factors, high expression of osmotically responsive genes, *HOS9* and *HOS10*, which are required for basal freezing tolerance [108, 109]. Similarly, microarray analysis led to the identification of the cold-stress-inducible *AP2* fam- ily transcription factor gene related to *ABI3/VP1* (*RAV1*) [57, 60] that might regulate plant growth under cold stress. *RAV1* is down regulated by epibrassinolide and transgenic *Arabi- dopsis* overexpressing *RAV1* exhibits a retardation of lateral root and rosette-leaf development. Further, the importance of *CBF*-independent pathways is also supported by analysis of mutants that have increased freezing tolerance, for example, mutations in eskimo1 (*ESK1*), a protein of unknown func- tion, result in constitutive freezing tolerance. The *Eskimo1* mutation was first identified as a mutation conferring frost survival without an acclimation period [110]. Later on, Bouchabke-Coussa *et al.* [111] clearly showed that *ES- KIMO1* mutants are more tolerant to freezing but only after acclimation. The genes that are affected by the *ESK1* muta- tion are distinct from those of the *CBF* regulon [112]. Recently, a *CBF* 1 gene has been introduced into tomato that resulted in transgenic plants showing higher activity of su- peroxide dismutase (SOD), higher non-photochemical quenching (NPQ), and lower malondialdehyde (MDA) con- tent. Thus suggesting that *CBF1* protein plays an important role in protection of PSII and PSI during low temperature stress at low irradiance [113].

Aforementioned studies clearly demonstrate the complex and interactive relationships among different pathways regulated by cold acclimation. Hence, elucidation of the mechanism regulating cold-regulated genes is an important goal in achieving a full understanding of cold acclimation. With the advent of molecular genetics and biotechnology, it is now possible to genetically engineer plants to be more tolerant to low temperature. The technologies can be so developed/ modified that will significantly improve breeding efficiency, resulting in rapid and accurate incorporation of cold tolerant genes into crop plants.

6. PROSPECTS OF COLD TOLERANT GENES IN HILL AGRICULTURE

Cold is an environmental factor that adversely affects crop quality, productivity, geographical distribution and growing season of many plant species [76]. Most temperate plants can acquire tolerance to freezing temperature by cold acclimatization, a process of prior exposure to low non-freezing temperature [44, 114, 115]. Plants of tropical and subtropical origins are sensitive to chilling temperature (0 °C-10 °C) and are incapable of cold acclimation. The identification of number of genes responsible for abiotic stress tolerance is of great importance not only for a better understanding of stress responses, but also for promising future crop improvement. Genes induced by stress includes the genes encoding transcription factors and proteins involved directly in response mechanisms. Variations in the expression of regulators such as transcription factors could lead to a protective status before the emergence of stress while genes involved in response mechanisms could be new targets for the improvement of plant plasticity and adaptive responses to stress [116]. It has repeatedly been speculated that certain *COR* (cold regulated) genes might have role in freezing tolerance. The continuous efforts have led to the identification of a number of genes such as the *COR 15a KINI, LTI 78, fad 7 etc.* of *A. thaliana* and their successful transfer to many crops by genetic engineering lead to the development of stress tolerant crop plants despite the complexity of plant responses to environmental stress [12, 66].

These expression of these COR genes are found to be critical in plants for both chilling tolerance [12, 52] and cold acclimation [13, 14, 76, 117]. They encode diverse array of proteins such as enzymes involved in respiration and metabolism of carbohydrates, lipids, phenylpropanoids and antioxidants: molecular chaperones, antifreeze proteins [44, 76, 118]. STIFDB (Stress responsive Transcription Factor Database), a database of stress related genes provides a platform to understand the stress-regulome of abiotic stress responsive genes in plants. The generic trends of the genes and transcription factors are available in STIFDB (available at http://caps.ncbs.res.in/stifdb). STIFDB will be a highly useful resource for a researcher working on abiotic stress responses in plants [119].

Extreme low temperature stress causes cause significant yield reductions, less extreme stress does not result in yield loses; it often results in crop quality reduction. Each year, approximately $ 2 Billion losses in crop production occurs due to low temperature damage worldwide. Some of the major losses include the 1995 early fall frosts in the US which caused losses of over $1 billion to corn and soybeans. California sustained $650M of damage in 1998 to the citrus crop due to a winter freeze, occasional freezes in Florida have shifted the citrus belt further south.

Crops that are more resistant to freezing stress would allow some geographical regions to grow more profitable and productive crops with less environmental risks. However despite continued efforts, traditional breeding has had only limited success in imparting crop plants with better freezing tolerance due to very little was known about the mechanisms that regulate chilling and freezing tolerance. With the advent of molecular genetics and biotechnolog y, it is now possible to genetically engineer plants to be more tolerant to many environmental adversities, including low temperature. Molecular studies have shown that several genes with various functions are induced by environmental stresses such as drought, high-salinity and low temperature in plants. Most of the dehydration responsive genes are induced by the plant hormone abscisic acid (ABA), but others are not. Expression analyses of dehydration-responsive genes have provided at least four independent regulatory systems (regulons) for gene expression in a model plant *Arabidopsis thaliana*. The *cis*-acting elements in the promoters of some genes that have a typical stress-inducible expression profile and the transcription factors that affect the expression of these genes have been analyzed [120].

7. FUTURE OUTLOOK AND CONCLUSION

Genetic engineering seems to be a faster and more viable option than conventional or molecular breeding for the development of improved crop varieties by the introduction and/or overexpression of selected genes. Also, it would be the only option when the source of genes of interest is a distant relatives, cross barrier species, or non-plant sources. Use of different genomic approaches and gene knockout strategies have made it possible to assess systematically and understand complex quantitative traits such as acquired tolerance to temperature extremes. A number of relevant low temperature sensitive genes have been identified by using genetic and molecular approaches. The information so obtained is continually emerging to enrich the *CBF* cold-responsive pathway. Thus, the *CBF/DREB*1 act as activators to trigger the assembly of several components of the cold acclimation response and thus increase the plants tolerance to low temperatures after exposure to non-freezing conditions. The development of improved varieties against abiotic stress have been achieved in a number of crop plants using *DREB1/CBF*. Studies on the other transcription factors associated with stress response are in progress.

The results of the transcriptome study revealed that the adaptation of plant to low temperature is a highly complex process. To overcome this problem a transgenic approach to promoting cold tolerance has been widely adopted, with some success. For example, overexpression of *E. coli* choline oxidase or trehalose-6-phosphate synthase fused to trehalose-6- phosphate phosphatase in transgenic rice enhanced tolerance to both salt and cold by increasing the accumulation of two compatible solutes (glycinebetaine and trehalose). A large number of genes involved in low temperature tolerance identified in different studies have unknown function but this problem can be solved in near future with continuously increasing information on the low-temperature transcriptome, proteome and metabolome. This information will also help to understand the complex network of molecular changes that occur in crop plants for chilling and freezing tolerance. A strategy combining the molecular, physiological and metabolic aspects of cold stress tolerance is required to understand the short- and long-term effects of the genes and their products, and molecular or cellular expression of the genes and the whole plant phenotype under stress. Collaborative research with many research groups to improve stress tolerant crop plants utilizing regulon biotechnology were undertaken under the aegis of CGAIR. We hope the results of these collaborative studies will contribute to the sustainable food production in developing countries and help to prevent the global-scale environmental damage.

CONFLICT OF INTEREST

The author confirms that he has no conflict of interest to declare for this publication.

ACKNOWLEDGEMENTS

Declared None.

DISCLOSURE

This is an updated chapter of the earlier review article [by Sanghera *et al.* "Engineering Cold Stress Tolerance in Crop Plants (review)". *Curr. Genomics*, 2011, 12: 30-43], with additional text and references.

REFERENCES

[1] Xin Z, Browse J. Cold comfort farm: the acclimation of plants to freezing temperatures. Plant CellEnviron 2001; 23: 893–902.

[2] Janmohammadi M, Tavkol-Afshari R, Mahfuzi S, Alizadeh H, Mock HP. Study of changes of cold-induced proteins during transition of vegetative to reproductive stages in winter wheat under field conditions. Iranian Journal of Field Crop Science 2012; 43: 93–111.

[3] Levitt J. Responses of plants to environmental stress In: Chilling, Freezing, and High Temperature Stress. Vol. 1. New York: Academic Press1980; pp. 497

[4] Shinozaki K, Yamaguchi-Shinozaki K. Molecular response to drought and cold stress. Curr OpinPlant Biol 1996; 7: 161–67.

[5] Thomashow MF. Role of cold-responsive genes in plant freezing tolerance. PlantPhysiol 1998; 118: 1–8.

[6] Gilmour SJ, Sebolt AM, Salazar MP, Everard JD, Thomashow MF. Overexpression of the *Arabidopsis* CBF3 transcriptional activator mimics multiple biochemical changes associated with cold acclimation. Plant Physiol 2000; 124: 1854–65.

[7] Yamaguchi-Shinozaki K, Shinozaki K. Transcriptional regulatory networks in cellular responses andtolerance to dehydration and cold stresses. Annu Rev Plant Biol 2006; 57: 781–803.

[8] Hannah MA, Heyer A G, Hincha DK. A global survey of gene regulation during cold acclimation in*Arabidopsis* thaliana. PLoS Genet 2005; 1: e26.

[9] Wani SH, Sandhu JS, Gosal SS. Genetic engineering of crop plants for abiotic stress tolerance. In: Malik CP, Kaur B, Wadhwani C, Eds. Advanced Topics in Plant Biotechnology and PlantBiology. New Delhi: MD Publications 2008; pp. 149–83.

[10] Gosal SS, Wani SH, Kang MS. Biotechnology and drought tolerance. J Crop Improv 2009; 23: 19–54.

[11] Wani SH, Gosal SS. Introduction of OsglyII gene into Indica rice through particle bombardment for increased salinity tolerance. Biol Plant 2011; 55(3): 536-40.

[12] Hsieh TH, Lee JT, Yang PT, *et al.* Heterology expression of the*Arabidopsis* C-repeat/dehydration response element binding factor 1 gene confers elevated tolerance to chilling and oxidative stresses in transgenic tomato. Plant Physiol 2002; 129: 1086–94.

[13] Knight H, Veale EL, Warren GJ, Knight MR. The sfr6 mutation in *Arabidopsis* suppresses lowtemperature induction of genes dependent on the CRT/DRE sequence motif. Plant Cell 1999; 11: 875–86.

[14] Tamminen I, Mäkelä P, Heino P, Palva ET. Ectopic expression of *ABI3* gene enhances freezing tolerance in response to abscisic acid and low temperature in *Arabidopsis thaliana*. PlantJ 2001; 25: 1–8.

[15] Ashraf M, Athar HR, Harris PJC, Kwon TR. Some prospective strategies for improving crop salt tolerance. Adv Agron 2008; 97: 45–110.

[16] Sanghera GS, Wani SH. Innovative approaches to enhance genetic potential of rice for higher productivity under temperate conditions of Kashmir. J. Plant Sci. Res. 2008; 24: 99–113.

[17] Rice environments In: Rice Almanac. 2nd. Philippines: IRRI; 1997. Anonymous. Rice environments; pp. 17–25.

[18] Sanghera GS, Zarger MA, Anwar A, *et al.* Studies on spikelet fertility and incidence of leaf blast on certain IRCTN rice genotypes under temperate conditions. Proceeding of National Symposium on plant Protection Strategies for sustainable Agri-Horticulture, SKAUST-Jammu; 2001Oct 12-13; India. SKAUST-Jammu; p. 125.

[19] Glaszmann JC, Kaw RN, Khush GS. Genetic divergence among cold tolerant rices (*Oryzasatva* L.) Euphytica 1990; 45: 95–104.

[20] Rasolofo PR. Screening rice cultivars at seedling stage and anthesis for low temperature tolerance inMadagascar. IRRN. 1986; 11: 12–13.

[21] Liu W, Yu K, He T, *et al.* The Low Temperature Induced Physiological Responses of Avena nuda L., a Cold-Tolerant Plant Species.The Scientific World Journal 2013; 1-7. doi: 10.1155/2013/658793.

[22] Janmohammadi M. Study of interrelationship between vegetative/reproductive transition stage and cold induced proteins expression using proteomics analysis in wheat grown under fieldconditions. PhD dissertation. Tehran: University of Tehran 2010.

[23] Singh BD. Plant Breeding. Kalyani Publishers 2000; pp. 443–60.

[24] Goulas E, Schubert M, Kieselbach T, *et al.* The chloroplast lumen and stromal proteomes of Arabidopsis thaliana show differential sensitivity toshort- and long-term exposure to low temperature. Plant Journal 2006; 47: 720–34.

[25] Janmohammadi M. Metabolomic analysis of low temperature responses in plants. Current Opinion inAgriculture Curr Opin Agric 2012; 1(1): 1–6.

[26] Saini JP, Tandon JP. Evaluating rices for cold tolerance. IRRN. 1985; 10: 9–10.

[27] Lia XH, Dai LY, Zhang JH. Factors related to cold tolerance at booting stage in *Oryza sativa* L. Chinese J Rice Sci 1998; 12: 6–10.

[28] Oliver SN, Van Dongen JT, Alfred SC, *et al.* Cold induced repression of the rice anther-specific cell wall invertase gene OSINV4 is correlated with sucrose accumulation and pollen sterility. Plant Cell Environ 2005; 28: 1534–51.

[29] Oliver SN, Dennis ES, Dolferus R. ABA regulates apoplastic sugar transport and is a potential signal for cold induced pollen sterility in rice. Plant Cell Physiol 2007; 48: 1319–30.

[30] Zhao X, Dolferuf R, Darvey N. Precision breeding of cold tolerant rice. IREC FarmersNewsletter 2008; 177: 12–13.

[31] Bhatnagar-Mathur P, Vadez V, Sharma KK. Transgenic approaches for abiotic stress tolerance in plants: retrospect and prospects. Plant Cell Rep. 2008; 27: 411–24.

[32] Steponkus PL. Role of the plasma membrane in freezing injury and cold acclimation. Annu. Rev.Plant Physiol. 1984; 35: 543–84.

[33] Steponkus PL, Uemura M, Webb MS. A contrast of the cryostability of the plasma membrane of winter rye and spring oat-two species that widely differ in their freezing tolerance and plasmamembrane lipid composition. In: Steponkus P L, editor. Adv.Low-Temperature Biol. Vol. 2. London: JAI Press1993; pp. 211–312.

[34] Horvath I, Glatz A, Varvasovszki V, *et al.* Membrane physical state controls the signaling mechanism of the heat shock response in Synechocystis PCC 6803: identification of hsp17 as a 'fluidity gene'. Proc Natl Acad Sci USA1998; 95: 3513–18.

[35] Orvar BL, Sangwan V, Omann F, Dhindsa RS. Early steps in cold sensing by plant cells: the role ofactin cytoskeleton and membrane fluidity. Plant J 2000; 23: 785–94.

[36] Wu J, Lightner J, Waewick N, Browse J. Low-temperature damage and subsequent recovery of fab1 mutant *Arabidopsis* exposed to 28°C. Plant Physiol 1997; 113: 347–56.

[37] Routaboul JM, Fischer SF, Browse J. Trienoic fatty acids are required to maintain chloroplast function at low temperatures. Plant Physiol 2000; 124: 1697–1705.

[38] Hugly S, Somerville CA. role for membrane lipid polyunsaturation in chloroplast biogenesis at lowtemperature. Plant Physiol 1992; 99: 197–202.

[39] Gomes E, Jakobsen MK, Axelsen KB, Geisler M, Palmgreen MG. Chilling tolerance in *Arabidopsis*involves ALA1, a member of a new family of putative aminophospholipidtranslocases. Plant Cell 2000; 12: 2441–53.

[40] Strauss G, Hauser H. Stabilization of lipid bilayer vesicles by sucrose during freezing. Proc NatlAcad Sci USA 1986; 83: 2422–26.

[41] Anchordoguy TJ, Rudolph AS, Carpenter JF, Crowe JH. Modes of interaction of cryoprotectants with membrane phospholipids during freezing. Cryobiology 1987; 24: 324–31.

[42] Epand RM, Shai Y, Segrest JP, Anantharamaiah GM. Mechanisms for the modulation of membranebilayer properties by amphipathic helical peptides. Biopolymers 1995; 37: 319–38.

[43] Olien CR, Smith MN. Ice adhesions in relation to freeze stress. Plant Physiol 1977; 60: 499–503.

[44] Guy CL. Cold acclimation and freezing stress tolerance: role of protein metabolism. Annu. Rev. PlantPhysiol. Plant Mol Biol 1990; 41: 187–223.

[45] Aroca R, Irigoyen JJ, Sanchez-diaz M. Drought enhances maize chilling tolerance. II. Photosynthetic traits and protective mechanisms against oxidative stress. Physiol. Plant. 2003; 117: 540–49.

[46] Livingston DP, Henson CA. Apoplastic sugars, fructans, fructan exohydrolase, and invertase in winter oat: responses to second-phase cold hardening. Plant Physiol 1998; 116: 403–08.

[47] Jones PG, Inouye M. The cold-shock response-a hot topic. Mol. Microbiol 1994; 11: 811–18.

[48] Nakaminami K, Karlson DT, Imai R. Functional conservation of cold shock domains in bacteria and higher plants. Proc Natl Acad Sci USA 2006; 103: 10122–27.

[49] Kim JS, Park SJ, Kwak KJ, Kim YO, Kim JY, Song J, *et al.* Cold shock domainproteins and glycine-rich RNA-binding proteins from *Arabidopsis* thaliana can promote the cold adaptation process in *Escherichia coli.* Nucleic Acids Res 2007; 35: 506–16.

[50] Kim YO, Kim JS, Kang H. Cold-inducible zinc finger-containing glycine-rich RNA-binding proteincontributes to the enhancement of freezing tolerance in *Arabidopsis* thaliana. Plant J 2005; 42: 890–900.

[51] Yu E, Owttrim GW. Characterization of the cold stress-induced cyanobacterial DEAD-box proteinCrhC as an RNA helicase. Nucleic Acids Res 2000; 28: 3926–34.

[52] Gong Z, Lee H, Xiong L, Jagendorf A, Stevenson B, Zhu JK. RNA helicase-like protein as an early regulator of transcription factors for plant chilling and freezing tolerance. Proc Natl Acad Sci USA 2002; 99: 11507–12.

[53] Gong Z, Dong CH, Lee H, *et al.* A DEAD box RNAhelicase is essential for mRNA export and important for development and stress responses in *Arabidopsis*.Plant Cell 2005; 17: 256–67.

[54] Dong CH, Hu X, Tang W, *et al.* A putative *Arabidopsis* nucleoporin, AtNUP160, is critical for RNA export and required for plant tolerance to cold stress. Mol Cell Biol 2006; 26: 9533–43.

[55] Snape JW, Semokhoskii A, Fish Sarma, R.N. Quarrie, S.A. Galiba G, Sutka J. Mapping frost tolerance loci in wheat and comparative mapping with other cereals. Acta Agron Hungar1997; 45: 265–70.

[56] Cushman JC, Bohnert HJ. Genomic approaches to plant stress tolerance. Curr Opin PlantBiol 2000; 3: 117–24.

[57] Flower S, Thomashow MF. *Arabidopsis* transcriptome profiling indicates that multiple regulatory pathways are activated during cold acclimation in addition to the CBF cold response pathway. PlantCell.2002; 14: 1675–90.

[58] Kreps JA, Wu Y, Chang HS, Zhu T, Wang X, Harper JF. Transcriptome changes for *Arabidopsis* in response to salt, osmotic, and cold stress. Plant Physiol 2002; 130: 2129–41.

[59] Seki M, Narusaka M, Ishida AJ, *et al.* Monitoring the expression profiles of 7000 *Arabidopsis* genes under drought, cold and high-salinity stresses using a full-length cDNA microarray. Plant J 2002; 3: 279–92.

[60] Vogel JT, Zarka DG, Van Anbuskirk HA, Fowler SG, Thomashow MF. Roles of the CBF2 andZAT12 transcription factors in configuring the low temperature transcriptome of *Arabidopsis*. PlantJ 2005; 41: 195–211.

[61] Shinozaki K, Yamaguchi-Shinozaki K, Seki M. Regulatory network of gene expression in the drought and cold stress responses. Curr Opin Plant Biol 2003; 6: 410–17.

[62] Seki M, Satou M, Sakurai T, *et al.* RIKEN *Arabidopsis* full length (RAFL) cDNA and its applications for expression profiling under abiotic stress conditions. J Exp Bot 2004; 55: 213–23.

[63] Holmberg N, Bulow L. Improving stress tolerance in plants by gene transfer. Trends PlantSci1998; 3: 61–66.

[64] Seki M, Kamei A, Yamaguchi-Shinozaki K, Shinozaki K. Molecular responses to drought, salinity and frost: common and different paths for plant protection. Curr Opin Plant Biol 2003; 14: 194–99.

[65] Tester M, Bacic A. Abiotic stress tolerance in grasses. From model plants to crop plants. PlantPhysiol 2005; 137: 791–93.

[66] Zhang JZ, Creelman RA, Zhu JK. From laboratory to field. Using information from *Arabidopsis* to engineer salt, cold, and drought tolerance in crops. Plant Physiol 2004; 135: 615–21.

[67] Vinocur B, Altman A. Recent advances in engineering plant tolerance to abiotic stress: achievements and limitations. Curr Opin Plant Biol 2005; 16: 123–32.

[68] Thomashow MF. So what's new in the field of plant cold acclimation? Lots! PlantPhysiol 2001; 125: 89–93.

[69] Artus NN, Uemura M, Steponkus PL, Gilmour SJ, Lin C, Thomashow MF. Constitutive expression of the cold-regulated *Arabidopsis* thaliana COR15a gene affects both choroplast and protoplast freezingtolerance. Proc Natl Acad Sci USA 1996; 93: 13404–09.

[70] Kodama H, Hamada T, Horiguchi G, Nishimura M, Iba K. Genetic enhancement of cold tolerance by expression of a gene for chloroplast *w*-3 fatty acid desaturase in transgenic tobacco. PlantPhysiol 1994; 105: 601–05.

[71] Ishizaki-Nishizawa O, Fuji T, Azume M, *et al.* Low- temperature resistance of higher plants is significantly enhanced by a nonspecific cyanobacterialdesaturase. Nat Biotechnol 1996; 14: 1003–06.

[72] Ariizumi T, Kishitani S, Inatsugi R, Nishida I, Murata N, Toriyama K. An increase in unsaturation of fatty acids in phosphatidylglycerol from leaves improves the rates of photosynthesis and growth at low temperatures in transgenic rice seedlings. Plant Cell Physiol 2002; 43: 751–58.
[73] Sakamoto A, Sulpice R, Hou CX, *et al*. Genetic modification of the fatty acid unsaturation of phosphatidylglycerol in chloroplasts alters the sensitivity of tobacco plants to cold stress. Plant Cell Environ 2003; 27: 99–105.
[74] Sui N, Li M, Zhao SJ, Li F, Liang H, Meng QW. Overexpression of glycerol-3-phosphate acyltransferase gene improves chilling tolerance in tomato. Planta 2007; 226: 1097–1108.
[75] Kumar N, Bhatt RP. Transgenics: An emerging approach for cold tolerance to enhance vegetables production in high altitude areas. Indian J Crop Sci 2006; 1: 8–12.
[76] Thomashow MF. Plant cold acclimation: freezing tolerance genes and regulatory mechanisms. AnnuRev Plant Physiol Plant Mol Biol 1999; 50: 571–99.
[77] Houde M, Dallaire S, N'dong D, Sarhan F. Overexpression of the acidic dehydrin WCOR410 improves freezing tolerance in transgenic strawberry leaves. Plant Biotechnol J 2004; 2: 381–87.
[78] Stockinger EJ, Gilmour SJ, Thomashow MF. *Arabidopsis* thaliana CBF1 encodes an AP2 domain-containing transcriptional activator that binds to the C-repeat/DRE, a cis-acting DNA regulatoryelement that stimulates transcription in response to low temperature and water deficit. Proc Natl Acad Sci USA1997; 94: 1035–40.
[79] Gilmour SJ, Zarka DG, Stockinger EJ, Salazar MP, Houghton JM, Thomashow MF. Low temperatureregulation of the *Arabidopsis* CBF family of AP2 transcriptional activators as an early step in cold- induced COR gene expression. Plant J 1998; 16: 433–42.
[80] Kasuga M, Liu Q, Miura S, Yamaguchi-Shinozaki K, Shinozaki K. Improving plant drought, salt, andfreezing tolerance by gene transfer of a single stress-inducible transcription factor. Nat Biotechnol1999; 17: 287–91.
[81] Jaglo-Ottosen KR, Gilmour SJ, Zarka DG, Schabenberger O, Thomashow MF. *Arabidopsis* CBF1overexpression induces COR genes and enhances freezing tolerance. Science 1998; 280: 104–06.
[82] Jaglo-Ottosen KR, Kleff S, Amundsen KL, *et al*. Components of the *Arabidopsis* C repeat/dehydration-responsive element binding factor cold- response pathway are conserved in *Brassica napus* and other plant species. Plant Physiol 2001; 127: 910–17.
[83] Castiglioni P, Warner D, Bensen RJ, Anstrom DC, Harrison J, Stoecker M, Abad M, Kumar G,Salvador S, D'Ordine R, Navarro S, Back S, Fernandes M, Targolli J, Dasgupta S, Bonin C, LuethyM H, Heard JE. Bacterial RNA Chaperones Confer Abiotic Stress Tolerance in Plants and ImprovedGrain Yield in Maize under Water-Limited Conditions. Plant Physiol 2008; 147: 446–55.
[84] Hajela RK, Horvath DP, Gilmour SJ, Thomashow MF. Molecular cloning and expression of *cor* (cold-regulated) genes in *Arabidopsis thaliana*. Plant Physiol 1990; 93: 1246–52.
[85] Gilmour SJ, Fowler SG, Thomashow MF. *Arabidopsis* transcriptional activators CBF1, CBF2, andCBF3 have matching functional activities. Plant Mol Biol 2004; 54: 767–81.
[86] Liu Q, Ksauga M, Sakuma Y, Abe H, Miura S, Yamaguchi-Shinozaki K, Shinozaki K. Two transcription factors, DREB1 and DREB2, with an EREBP/AP2 DNA binding domain separate twocellular signal transduction pathways in drought and low-temperature-responsive gene expression, respectively, in Arabidopsis. Plant Cell 1998; 10: 1391–1406.
[87] Haake V, Cook D, Riechmann JL, Pineda O, Thomashow MF, Zhang JZ. Transcription factor CBF4is a regulator of drought adaptation in *Arabidopsis*. Plant Physiol 2002; 130: 639–48.
[88] Kasuga M, Miura S, Shinozaki K, Yamaguchi-Shinozaki K. A combination of the *Arabidopsis*DREB1A gene and stress-inducible rd29A promoter improved drought and low-temperature stress tolerance in tobacco by gene transfer. Plant Cell Physiol 2004; 45: 346–50.
[89] Pellegrineschi A, Reynolds M, Pacheco M, Brito RM, Almeraya R, Yamaguch-Shinozaki K, Hoisington D. Stress-induced expression in wheat of the *Arabidopsis* thaliana DREB1A gene delayswater stress symptoms under greenhouse conditions. Genome. 2004; 47: 493–500.
[90] Agarwal M, Hao Y, Kapoor A, Dong CH, Fujii H, Zheng X, Zhu JK. A R2R3 type MYBtranscription factor is involved in the cold regulation of CBF genes and in acquired freezing tolerance. J Biol Chem 2006; 281: 37636–45.
[91] Murata N, Ishizaki-Nishizawa O, Higashi S, Hayashi S, Tasaka Y, Nishida I. Genetically engineered alteration in the chilling sensitivity of plants. Nature 1992; 356: 710–13.

[92] Kovtum, Chiu WL, Tena G, Sheen J. Functional analysis of oxidative stress-activated mitogen activated protein kinase cascade in plants. Proc Natl Acad Sci USA 2000; 97: 2940–3005.
[93] Pennycooke JC, Jones ML, Stushnoff C. Down-regulating α -Galactosidase enhances freezing tolerance in transgenic Petunia. Plant Physiol 2003; 133: 901–09.
[94] Hara M, Terashima S, Fukaya T, Kubol T. Enhancement of cold tolerance and inhibition of lipid peroxidation by citrus dehydrin in transgenic tobacco. Planta 2003; 217: 290–98.
[95] Puhakainen T, Hess MW, Makela P, Svensson J, Heino P, Palva ET. Overexpression of multipledehydrin genes enhances tolerance to freezing stress in *Arabidopsis*. Plant Mol Biol 2004; 54: 743–53.
[96] Kaye C, Neven L, Hofig A, Li QB, Haskell D, Guy C. Characterization of a gene for spinach CAP160 and expression of two spinach cold-acclimation proteins in tobacco. Plant Physiol 1998; 116: 1367–77.
[97] Iturriaga G, Schneider K, Salamini F, Bartels D. Expression of desiccation-related proteins from theresurrection plant Craterostigma plantagineum in transgenic tobacco. Plant Mol Biol 1992; 20: 555–58.
[98] Dai X, Xu Y, Ma Q, Xu W, Wang T, Xue Y, Chong K. Overexpression of an R1R2R3 MYB gene, OsMYB3R-2, increases tolerance to freezing, drought and salt stress in transgenic Arabidopsis. Plant Physiol 2007; 143: 1739–51.
[99] Pramanik MH, Imai R. Functional identification of a trehalose 6-phosphate phosphatase gene that isinvolved in transient induction of trehalose biosynthesis during chilling stress in rice. Plant Mol Biol 2005; 58: 751–62.
[100] Garg AK JK, Kim TG, Owens AP, Ranwala YD, Choi LV, Kochian RJ, Wu R. Trehaloseaccumulation in rice plants confers high tolerance levels to different abiotic stresses. Proc Natl Acad Sci 2002; 99: 15898–903..
[101] Jang IC, Oh SJ, Seo JS, Choi WB, Song SI, Kim CH, Kim YS, Seo HS, Choi YD, Nahm BH.Expression of a bi-functional fusion of the *Escherichia coli* genes for trehalose-6-phosphate synthase and trehalose- 6-phosphate phosphatase in transgenic rice plants increases trehalose accumulation andabiotic stress tolerance without stunting growth. Plant Physiol 2003; 131: 516–24.
[102] Ge LF, Chao DY, Shi M, Zhu MZ, Gao JP, Lin HX. Overexpression of the trehalose-6-phosphate phosphatase gene OsTPP1 confers stress tolerance in rice and results in the activation of stressresponsive genes. Planta 2008; 228: 191–201.
[103] Iordachescu M, Imai R. Trehalose biosynthesis in response to abiotic stresses. J. Integr PlantBiol 2008; 50: 1223–29.
[104] Su CF, Wang YC, Hsieh TH, Lu CA, Tseng TH, Yu SM. A novel MYBS3-dependent pathway confers cold tolerance in rice. Plant Physiol 2010; 153: 145–58.
[105] Shinozaki K, Yamaguchi-Shinozaki K. Molecular responses to dehydration and low temperature: differences and cross-talk between two stress signaling pathways. Curr Opin PlantBiol 2000; 3: 217–23.
[106] Achard P, Gong F, Cheminant S, Alioua M, Hedden P, Genschika P. The cold-inducible CBF1 factor-dependent signaling pathway modulates the accumulation of the growth-repressing DELLAproteins *via* its effect on gibberellin metabolism. Plant Cell 2008; 20: 2117–29.
[107] Kim JC, Lee SH, Cheong YH, Yoo CM, Lee SI, Chun HJ, Yun DJ, Hong JC, Lee SY, Lim CO, ChoMJ. A novel cold-inducible zinc finger protein from soybean, SCOF-1, enhances cold tolerance in transgenic plants. Plant J 2001; 25: 247–59.
[108] Zhu J, Shi H, Lee BH, Damsz B, Cheng S, Strim V, Zhu JK, Hasegawa PM, Bressan RA.An*Arabidopsis* homeodomain transcription factor gene, *HOS9*, mediates cold tolerance through aCBF-independent pathway. Proc Natl Acad Sci USA 2004; 101: 9873–78.
[109] Zhu J, Verslues PE, Zheng X, *et al. HOS10* encodes an R2R3-type MYB transcription factor essential for cold acclimation in plants. Proc Natl Acad Sci USA 2005; 102: 9966–71.
[110] Xin Z, Browse J. Eskimo1 mutants of *Arabidopsis* are constitutively freezing-tolerant. Proc NatlAcad Sci USA 1998; 95: 7799–804.
[111] Bouchabke-Coussa O, Quashie M, Seoane-Redondo J, Fortabat M, Gery C, Yu A, Linderme D, Trouverie J, Granier F, Téoulé E, Durand-Tardif M. *ESKIMO1* is a key gene involved in watereconomy as well as cold acclimation and salt tolerance. BMC Plant Biol 2008; 8: 125–51.
[112] Xin Z, Mandaokar A, Chen J, Last RL, Browse J. Arabidopsis *ESK1* encodes a novel regulator of freezing tolerance. Plant J 2007; 49: 786–99.
[113] Zhang YJ, Yang JS, Guo SJ, Meng JJ, Zhang L, Wan SB, He QW, Li XG. Over-expression of the*Arabidopsis* CBF1 gene improves resistance of tomato leaves to low temperature under lowirradiance. Plant Biol 2010; 13(2): 362-7.

[114] Hughes MA, Dunn MA. The molecular biology of plant acclimation to low temperature. J ExpBot 1996; 47: 291-305.
[115] Browse J, Xin Z. Temperature sensing and cold acclimation. Curr Opin Plant Biol 2001; 4: 241–46.
[116] Umezawa T, Fujita M, Fujita Y, Yamaguchi-Shinozaki K, Shinozaki K. Engineering droughttolerance in plants: discovering and tailoring genes to unlock the future. Curr OpinBiotechnol 2006; 17: 113–22.
[117] Chinnusamy V, Zhu J, Zhu JK. Gene regulation during cold acclimation in plants. PhysiolPlant 2006; 126: 52–61.
[118] Mahapatra SS, Wolfraim L, Poole RJ, Dhindsa RS. Molecular cloning and relationship to freezing tolerance of cold-acclimation-specific genes of alfalfa. Plant Physiol 1989; 89: 375–80.
[119] Shameer K, Ambika S, Varghese SM, Karaba N, Udayakumar M, Sowdhamini R. STIFDB-*Arabidopsis* stress responsive transcription factor database. Inter J Plant Genom 2009; 5: 1–8.
[120] Nakashima K, Yamaguchi-Shinozaki K. Molecular studies on stress-Responsive gene expression in*Arabidopsis* and improvement of stress tolerance in crop plants by regulonbiotechnology. JARQ 2005; 39: 221–29.
[121] Sen Gupta A, Heinen JL, Holady AS, Burke JJ, Allen RD. Increased resistance to oxidative stress in transgenic plants that over-express chloroplastic Cu/Zn superoxide dismutase. Proc Natl Acad SciUSA1993; 90: 1629–33.
[122] Pilon-Smits EAH, Ebskamp MJM, Paul MJ, Jeuken MJW, Weisbeek P J, Smeekens S.C.M. Improved performance of transgenic fructan-accumulating tobacco under drought stress. PlantPhysiol 1995; 107: 125–30.
[123] Mckersie BD, Bowley SR, Harjanto E, Leprince O. Water deficit tolerance and field performance of transgenic alfalfa overexpressing superoxide dismutase. Plant Physiol 1996; 111: 1177–81.
[124] Roxas VP, Smith RK, Jr, Allen ER, Allen RD. Overexpression of glutathione S-transferase/glutathione peroxidase enhances the growth of transgenic tobacco seedlings during stress. Nat Biotechnol1997; 15: 988–91.
[125] Ouellet F, Vazquez-Tello A, Sarhan F. The wheat wcs120 promoter is cold-inducible in both monocotyledonous and dicotyledonous species. FEBS Lett 1998; 423: 324–28.
[126] Sakamoto A, Murata AN. Metabolic engineering of rice leading to biosynthesis of glycinebetaine andtolerance to salt and cold. Plant Mol Biol 1998; 38: 1011–19.
[127] Hayashi H, Sakamoto A, Nonaka H, Chen THH, Murata N. Enhanced germination under high-salt conditions of seeds of transgenic *Arabidopsis* with a bacterial gene (codA) for choline oxidase. JPlant Res1998; 111: 357–62.
[128] Nanjo T, Kobayashi M, Yoshiba Y, Kakubari Y, Yamaguchi-Shinozaki K, Shinozaki K. Antisense supression of proline degradation improves tolerance to freezing and salinityin *Arabidopsis* thaliana. FEBS Lett 1999; 461: 205–10.
[129] Lee SC, Won HK, An K, An G, Kim SR. Ectopic expression of a cold- inducible transcription factor, CBF1/DREB1b, in transgenic rice (*Oryza sativa* L.) Mol Cells 2004; 18: 107–14.
[130] Mukhopadhyay A, Vij S, Tyagi AK. Overexpression of a zinc-finger protein gene from rice confers tolerance to cold, dehydration, and salt stress in transgenic tobacco. Proc Natl Acad SciUSA 2004; 101: 6309–14.
[131] Vannini C, Locatelli F, Bracale M, Magnani E, Marsoni M, Osnato M, Mattana M, Baldoni E, Coraggio I. Overexpression of the rice Osmyb4 gene increases chilling and freezing toleranceof *Arabidopsis*thaliana plants. Plant J 2004; 37: 115–27.
[132] Zhu J, Verslues PE, Zheng X, *et al* . *HOS10* encodes an R2R3-type MYB transcription factor essential for cold acclimation in plants. Proc Natl Acad Sci USA 2005; 102: 9966–71.
[133] Nakayama K, Okawa K, Kakizaki T, Honma T, Itoh H, Inaba T. *Arabidopsis* Cor15am is a chloroplast stromal protein that has cryoprotective activity and forms oligomers. PlantPhysiol 2007; 144: 513–23.
[134] Chen QF, Xiao S, Chye ML. Overexpression of the *Arabidopsis* 10-kDa acyl-CoA-binding proteinACBP6 enhances freezing tolerance. Plant Physiol 2008; 148: 304–15.
[135] Ma Q, Dai X, Xu Y, *et al.* Enhanced tolerance to chilling stress in OSMYB3R-2 transgenic rice is mediated by alteration in cell cycle and ectopic expression of stress genes. Plant Physiol. 2009; 150: 244–56.
[136] Kim MH, Sasaki K, Imai R. Cold shock domain protein 3 regulates freezing tolerance in *Arabidopsis*thaliana. J Biol Chem 2009; 284: 23454–60.

[137] Zhai H, Bai X, Zhu Y, *et al.* A single-repeat R3-MYB transcription factor MYBC1 negatively regulates freezing tolerance in *Arabidopsis*. Biochem Biophys Res Commun 2010; 394: 1018–23.

[138] Zhu B, Xiong AS, Peng RH, Xu J, Jin XF, Memg XR, Quan-Hong Y. Over-expression of *ThpI* from*Choristoneura fumiferana* enhances tolerance to cold in *Arabidopsis*. Mol BiolRep 2010; 37: 961–66.

[139] Checker VG, Chhibbar AK, Khurana, P.Stress-inducible expression of barley Hva1 gene in transgenic mulberry displays enhanced tolerance against drought, salinity and cold stress. Transgenicresearch 2012; 21: 939-57.

[140] Patade VY, Khatri D, Kumari M, Grover A, Gupta SM, Ahmed Z. Cold tolerance inOsmotin transgenic tomato (*Solanum lycopersicum* L.) is associated with modulation in transcript abundance of stress responsive genes. Springer Plus 2013; 2 (1), 117.

[141] Miura K, Shiba H, Ohta M, *et al.* SlICE1 encoding a MYC-type transcription factor controls cold tolerance in tomato, *Solanum lycopersicum*. Plant Biotechnol 2012; 29: 253–60.

[142] Miura K, Sato A, Shiba, H, Kang, SW, Kamada H, Ezura, H. Accumulation of antioxidants and antioxidant activity in tomato, *Solanum lycopersicum*, are enhanced by the transcription factorSlICE1. Plant Biotechnol 2012; 29; 261–69.

[143] Hu Y, Zhang L, Zhao L, *et al.* Trichostatin A selectively suppresses the cold-induced transcription of the ZmDREB1 gene in maize. PLoS One 2011; 6: e22132. doi: 10.1371/journal.pone.0022132

[144] Fan W, Zhang M, Zhang H, Zhang P. Improved Tolerance to Various Abiotic Stresses inTransgenic Sweet Potato (*Ipomoea batatas*) Expressing Spinach Betaine Aldehyde Dehydrogenase. PLoS ONE 2012; 7(5): e37344. doi: 10.1371/journal.pone.0037344

[145] Faxiang Wan, Yu Pan, Jinghua Li, *et al.* Heterologous expression of Arabidopsis C -repeat binding factor 3 (AtCBF3) and cold -regulated 15A (AtCOR15A) enhanced chilling tolerance in transgenic eggplant (*Solanum melongena* L.). Plant Cell Rep 2014; 33: 1670 doi: 10.1007/s00299-014-1670-z.

[146] Shurong D, Jian S, Zhao R *et al.* Populus euphratica apyrase, PeAPY2, enhances cold tolerance by modulating vesicular trafficking and extracellular ATP in Arabidopsis plants. Plant Physiol 2015; pp.15.00581v1-pp.00581.2015.

Subject Index

D

E

F

O

P

R

S

T

U

V

W

X

Z

Made in the USA
Las Vegas, NV
03 November 2026

57424578R00128